DEVELOPMENTS IN MEAT SCIENCE—3

CONTENTS OF VOLUMES 1 AND 2

Volume 1

Volume 2

DEVELOPMENTS IN MEAT SCIENCE—3

Edited by

RALSTON LAWRIE

Professor of Food Science and Head,
Department of Applied Biochemistry and Food Science,
University of Nottingham, UK

ELSEVIER APPLIED SCIENCE PUBLISHERS
LONDON and NEW YORK

ELSEVIER APPLIED SCIENCE PUBLISHERS LTD
Crown House, Linton Road, Barking, Essex IG11 8JU, England

Sole Distributor in the USA and Canada
ELSEVIER SCIENCE PUBLISHING CO., INC.
52 Vanderbilt Avenue, New York, NY 10017, USA

British Library Cataloguing in Publication Data

Developments in meat science.—3
1. Meat industry and trade—Periodicals
664′.9′005 TS1960

ISBN 0-85334-361-6
ISSN 0144-8811

WITH 13 TABLES AND 13 ILLUSTRATIONS

The selection and presentation of material and the opinions expressed in this publication are the sole responsibility of the authors concerned.

Printed in Great Britain by Galliard (Printers) Ltd, Great Yarmouth

PREFACE

The sequence of chapters in the third volume of topics in the field of meat science again reflects the chronology of events in the history of the commodity. Chapter 1, jointly written by Professor Berg of Canada and Professor Butterfield of Australia, considers the potential for increasing the production of meat both by improvements in traditional meat animals and by exploitation of hitherto underdeveloped species.

Since 80 % of the dry weight of meat consists of protein, it is important to increase our understanding of its nature. The subtlety of the pattern of proteins in muscle becomes more apparent as investigations proceed. Professor Maruyama has reviewed recent findings on the structural proteins of muscle in Chapter 2.

The establishment of the European Economic Community has generated general interest in standardisation. Thus, it has been appreciated that the methods used to stun meat animals at slaughter vary significantly, and this has stimulated investigation of the effects of various stunning procedures. Mr T. M. Leach is an experienced veterinarian who is well qualified to assess this topic.

With the ever-increasing development of centralised preparation of pre-wrapped, consumer-portion meat cuts, it seems appropriate that Mr A. A. Taylor should draw attention, in Chapter 4, to the problems and benefits which are likely to arise.

Following the Second World War, much research effort was expended in examining the possibility of preserving foods by ionising radiation—a topic foreshadowed by the work of D. E. Lea about 60 years ago. Although effective, the use of ionising radiation was discouraged by fears that traces of carcinogens might be produced in foodstuffs. Long-term tests have now assured Public Health Authorities. Prolongation of the refrigerated storage life of meat by relatively low doses of ionising radiation is likely to be permitted. The chapter by Professor Elias considers the augury for this mode of preservation.

Tenderness in meat is a predominant desideratum, and the procedures for achieving it are many. Dr Macfarlane of CSIRO, Australia, has been one of the pioneers in the use of high pressure to tenderise the commodity. Although commercial application is perhaps some time in the future, it is deemed appropriate that the procedure should be described at this time.

The ecological consequences of environmental pollution have emphasised the problems of contamination generally and in food in particular. Meat can be the repository of various contaminants and, in the last chapter, Professor Ruiter has reviewed the nature and extent of contaminants in meat and meat products.

RALSTON LAWRIE

CONTENTS

LIST OF CONTRIBUTORS

R. T. BERG
Department of Animal Science, University of Alberta, Edmonton, Canada T6G 2P5

R. M. BUTTERFIELD
Department of Veterinary Science, University of Sydney, New South Wales 2006, Australia

P. S. ELIAS
Federal Research Centre for Nutrition, 75 Karlsruhe 1, Postfach 3640, Federal Republic of Germany

T. M. LEACH
School of Veterinary Studies, Murdoch University, Murdoch, Western Australia 6150, Australia

J. J. MACFARLANE
CSIRO Division of Food Research, Meat Research Laboratory, PO Box 12, Cannon Hill, Queensland 4170, Australia

KOSCAK MARUYAMA
Department of Biology, Faculty of Science, Chiba University, Yayoi-cho, Chiba 260, Japan

A. Ruiter
Faculty of Veterinary Medicine, University of Utrecht, PO Box 80.175, 3508 TD Utrecht, The Netherlands

A. A. Taylor
Agricultural and Food Research Council Food Research Institute, Langford, Bristol BS18 7DY, UK

Chapter 1

NEW AND IMPROVED TYPES OF MEAT ANIMALS

R. T. Berg

Department of Animal Science,
University of Alberta, Edmonton, Canada

and

R. M. Butterfield

Department of Veterinary Science,
University of Sydney, Australia

1. HISTORICAL PERSPECTIVES IN THE USE OF ANIMALS FOR MEAT

Many species of animals have been and are being used for food. In fact few animal species have escaped man's desire for meat. Hunting and gathering still plays a significant role as a source of meat in some primitive cultures but domestication has been the major contributor filling the large demand of modern societies for meat.

Domestication of animals goes back into antiquity. Current thought suggests that domestication was not a sudden invention of enlightened, advanced members of our evolving species, but that there have been from very early times (as in fact there still remain today) hunters, hunter-gatherers and those practising controlled hunting, nomadic herding, to close herding and use for draft riding and milk of a variety of species. Domestication seems to have involved a gradual development of man–animal symbiotic relationships in the economic exploitation of the habitat. There seem to be no clear cut differences between animal species which are or have been domesticated in the past and those which have not

Species were probably selected or abandoned on the basis of whether their food/habitat provided an efficient mode of exploitation under prevailing economic systems.[1]

Domestication of other species will probably depend more on the development of suitable husbandry techniques and the potential for exploitation of some available resources (e.g. underdeveloped areas) rather than on the basic response to man. As examples the present husbanding and exploitation of the red deer in New Zealand, the bison in North America and the water buffalo in Australia can be noted.

It must be appreciated that there are many species of animals which fall under the umbrella of 'meat animals' in specific regions of the world, but which are never eaten in others (e.g. horse, seal, polar bear, rhinoceros, hippopotamus, elephant, whale, possum). Lawrie[2] expressed the extent of 'meat animals' to embrace even human flesh. In fact, the preferences of some ethnic groups are abhorrent to others. We must recognise that some species could be large potential meat producers in areas where they are currently not eaten with only minor adjustments of consumer tastes. The kangaroo is such an example. The recent development of electric fencing to keep kangaroos off pasture in Australia[3] may well be turned around to keep kangaroos within areas where their productive capability could be fully expressed and their harvest achieved in a manner acceptable for human consumption. All that is needed is the appropriate development of demand and the right economic climate.

When domestication is viewed as economic exploitation involving man–animal symbiotic relationships, continuation of existing systems or development of new systems will depend on the flexibility and adaptability of the system to changing economic conditions. Human ingenuity within economic constraints will ultimately determine the continued husbandry of presently domesticated species and the further domestication of others.

New domesticated species will not probably equal present domestic animals in traditional assessment criteria for meat production, i.e. growth rate, production per hectare or a more favourable carcass composition. Hudson[4] in reviewing the potential productivity of world grazing systems, concluded that wildlife assemblages of East African savannahs did not support earlier optimism relating to productivity. The highest recorded biomass density was 200 kg/ha for Kiwa National Park, Zaire and the highest reported productivities were 10 kg/ha/year. Successful exploitation of such systems will depend on economic factors and probably not particularly on the inherent meat production qualities of the species exploited.

At some stage in domestication, selective breeding and modification of management become a part of the system. Both have the objective of improving the economics of the system, either by improving the product so it will command a better price or by reducing the cost of production. Domestic species used for meat production have been modified in meat production characteristics such as growth, development and carcass composition. An understanding of factors which influence growth and development allows one to mould a more desirable product. Much research has been done toward a better understanding of the growth process and this will be reviewed in a later section. Within the major meat producing species much potential still exists to modify them in a direction toward improved meat production. We believe such modifications will constitute the major effort of researchers in the near term. Newly domesticated species may, after exploitation systems become somewhat stabilised, pass through the same processes of improvement as the major domestic meat producing species. Which species become important in meat production will depend mostly on economic potentials rather than inherent superiority in meat production characteristics.

2. IMPORTANT CHARACTERISTICS OF MEAT ANIMALS

2.1. Production Traits

Any species husbanded for meat production must meet certain production criteria which contribute to economic viability of the enterprise. These include adaptability to the environment or habitat, reproduction, survival, longevity and efficiency. Adaptability, survival and longevity are under continuous pressure of natural selection as is reproduction to some extent. Natural selection favours survival and reproduction under the given conditions of the environment. The strategy for survival will probably not be the same as that for maximising production efficiency. Under stressful conditions survival (adaptability) will be in conflict with growth rate[5] and reproductive rate. Long term adaptability under such conditions will depend on the animal's capacity to reduce maintenance requirements and to curb reproduction during the most stressful parts of the cycle.

Modern animal husbandry has evolved management systems which reduce the impact of natural stressors (e.g. food shortage, extreme temperatures, diseases and parasites). In the absence or reduction of these stressors, selection for what are considered productive traits is more effective. Thus reproduction in major domestic species has been enhanced.

Under improved conditions, cattle achieve annual calving cycles and the incidence of twinning increases, sheep have been developed which produce litters and litter size in pigs has been considerably enhanced.

Improvement of a meat animal production enterprise thus involves both genetics and management—animals which are more adaptable and productive under the prevailing conditions are selected and management alters the effective environment to enhance productivity. Both processes have evolved together to result in viable meat animal production enterprises which are often quite unique and localised depending on a complex of interactions of the natural conditions and resources, management skills and market availability.

Efficiency in a broad sense will be important to sustained viability of a meat production enterprise. Efficiency as a concept has been discussed by Spedding.[6] Biological efficiency may be measured as output (liveweight produced) per hectare, per unit of feed, per unit of energy in feed or per unit of protein in the feed. Economic efficiency or profitability would be the bottom line for assessing the viability of an enterprise.

Methods have not been developed whereby economic efficiency can be assessed for individual animals for breeding and selection decisions. Therefore more easily measurable selection criteria which relate to economic efficiency have been developed. Attempts are generally made to improve reproductive performance up to a level compatible with feed resources and management skills. Reproduction has to be measured as animals surviving to market rather than as animals born, which brings into focus optimum reproduction for given conditions. Improving reproduction within limits imposed by the environment is generally considered to improve economic efficiency but it is recognised that costs associated with any improvement have to be considered.

Productive efficiency of individual meat animals could be assessed by liveweight gain per unit of feed. However feed inputs are expensive to measure. Liveweight gain in itself is a commonly used criterion to measure productive efficiency of individual animals. It is related to efficiency in that it reduces the period of maintenance for an animal destined for market at a given liveweight. Growth rate is also considered to be genetically correlated to the efficiency of feed conversion with faster growing animals tending to be more efficient. A drawback of liveweight gain as a measure of productive efficiency is that only part of the gain has economic value (e.g. the carcass and certain by-products). Further the carcass value is dependent on its composition, most critically its lean meat (muscle) yield. Thus Fowler *et al.*[7] have proposed lean carcass feed efficiency as the most reliable measure of individual productive efficiency in pigs and lean carcass growth rate as an alternative when individual feed inputs cannot be obtained. Both lean carcass feed efficiency and lean carcass growth rate depend on accurate methods of estimating carcass composition in a live animal. Methods for estimating live animal carcass composition have been steadily improving (reviewed by Berg and Walters[8]). The implications of selecting for growth rate will be discussed later.

2.2. Carcass Composition

As meat production systems become more sophisticated, critical assessment of carcass composition becomes more important. The most important tissue in meat animals is muscle which along with an acceptable level of fat becomes the edible portion of a carcass. Although an edible portion is the critical end point, meat animals are marketed in various ways, depending on the sophistication of the market, e.g. by the animal, by liveweight, by carcass weight, and by carcass weight and estimated edible product yield.

When animals are sold on a liveweight basis, dressing percentage is a critical unknown factor which must be estimated. When sale is by carcass weight, edible product must be estimated. Distribution of edible product (muscle and fat) is also important as different parts of the carcass have different value. An ideal carcass from a meat animal was described by Berg and Butterfield[9] as one which had a high muscle:bone ratio and an optimum level of fat. Berg and Butterfield[10] suggested that muscle as a proportion of liveweight might be a useful end point measurement for an animal intended for meat. It incorporates in one figure dressing percentage and muscle yield.

The major tissues of a carcass are muscle, fat and bone. To achieve the most desirable carcass composition which has a maximum amount of muscle, a minimum of bone and an optimum level of fat, it is necessary to understand how each of these tissues grow and develop and what effect external factors have on their growth and development. Also as tissues grow and develop it is advantageous to understand what factors influence the distribution of each of the tissues throughout the carcass in order that these processes might be controlled toward a goal of enhancing carcass value.

A general pattern for growth and development of muscle, fat and bone was described by Berg and Walters.[8] Bone is early developing, muscle is intermediate and fat is a late developing tissue in all meat animal species. Thus as an animal grows muscle:bone ratio increases to a plateau at maturity and fat:muscle ratio increases to maturity. The deposition of fat relative to growth of muscle and bone is influenced by the sex of the animal, by its breed type and by the plane of nutrition on which it is reared.[8] Heifers fatten earlier relative to liveweight than steers which in turn fatten earlier than bulls. In pigs the early fatteners are barrows followed by gilts and boars. In sheep the sexes follow the same pattern as for cattle.

In cattle, British beef breeds (Hereford, Shorthorn and Angus) are early fatteners and Continental breeds (Charolais, Simmental, Chianina) are late fatteners. Breed differences in propensity to fatten also exist for sheep and pigs.

Plane of nutrition has a well known effect on fattening pattern, with a high plane promoting early fattening and a low plane leading to delayed fattening. Thus some control of the proportion of fat in a carcass can be exercised by marketing at light weights, taking account of sex differences, choice of breed and by controlling energy intake.

The proportion of muscle in a carcass in the first instance varies inversely with the proportion of fat, so factors which increase proportion of fat will decrease the proportion of muscle and vice versa. Interestingly the proportion of muscle relative to liveweight seems to be little influenced by fatness.[10,11] Muscle relative to liveweight seems to be related to muscle:bone ratio. Muscle:bone ratio seems to increase predictably with weight within a genotype and considerable differences exist between various breed types. Nutritional plane does not seem to have a great effect on muscle:bone ratio at least when positive gains in weight are occurring.

Muscle-weight distribution, in that it may relate to the proportion of muscle in high-priced cuts, could influence carcass value. Muscle-weight distribution seems to follow functional demands and is quite predictable within a species at given stages of development.[10] Since Butterfield[12] first demonstrated similarity of muscle-weight distribution among breeds of cattle, considerable research has been done comparing breeds and types in cattle, sheep, pigs and other species (review by Berg and Walters).[8] Some statistically significant differences between breeds have been found but they were generally small and not considered commercially important. Bergstrom[13] and Taylor *et al.*[14] consider differences which do exist are probably genetic and could be exploited. Very heavily muscled breeds of pigs and cattle have been shown to have a more favourable muscle-weight distribution than more normally muscled breeds.[15,16]

Fat partitioning among the depots and its distribution throughout the carcass influences the commercial value of a carcass. Subcutaneous fat is later developing than is intermuscular fat while results for kidney and channel fat have been inconclusive.[8] Many reports indicate that dairy breeds deposit a higher proportion of fat internally and less subcutaneously than traditional beef breeds. It seems that fat partitioning among the depots has both genetic and environmental components. Both genetic and environmental tendencies to early fattening result in increased subcutaneous:intermuscular ratios while delayed fattening from whatever cause has an opposite effect.

With regard to fat distribution, growth gradients exist throughout the carcass with low growth coefficients in the distal limbs increasing centripetally on the limbs and still further toward the ribs and loin area. Highest growth coefficients for fat were found in flank and brisket areas.[8] Generally fat distribution in the major meat animal species studied is predictable within species by the total level of fatness achieved. No way to change fat distribution at a given level of fatness has been suggested to date.

3. SOME PRINCIPLES GOVERNING THE FUTURE ASSESSMENT OF IMPROVEMENT OF BODY COMPOSITION

Until recent times 'improvement' of body composition has been assessed mostly through subjective processes linked to eye appraisal of live animals. The modern trend is to assess body composition by objective means. However, to enable sensible comparisons of objective data a great deal of information must be obtained regarding the process of growth, and techniques must be evolved to determine the real value of comparisons made on different bases, such as age, liveweight, tissue weight, proportion of maturity and others.

Comparisons of meat animals to assess superiority in a particular market are relatively simple as the criteria of excellence of that market can be defined and the animals assessed relative to them. Unfortunately, 'markets' are ephemeral and such comparisons are, therefore, of fleeting value in the long term of animal breeding. For this reason it is necessary to understand the growth process throughout as much of the lifespan as possible, rather than to concentrate on the composition of animals in line with current market criteria if comparisons of lasting value are to be made. Taylor[17] stated, comparisons of carcass characteristics at fixed age or fixed weight 'have some economic significance but no biological justification'.

Before soundly-based decisions can be made on the use of any bases such as weight, age, fatness, stage of maturity, etc. for comparison of the proportion of a particular structure within the body it is necessary to understand how that structure grows relative to the total animal, or total tissue of which it is an intrinsic unit, if the results are to be acceptable. University of Sydney workers[18–25] have completed studies using sheep with the aim of developing principles which will allow more confident comparisons to be made of meat animals irrespective of their mature size or sex.

By the development of maturity coefficients it is possible to determine if tissues and structures follow similar or different paths to maturity in different strains and sexes. The form of the pathway to maturity of each structure determines the manner in which genetic comparisons of its proportion of the whole should be made. If a structure is earlier or later maturing than the whole animal then comparisons of animals of different mature size at the same liveweight give different answers to comparisons at the same proportion of mature liveweight. The comparison at equal liveweight contains a mixture of developmental difference and maturity difference whereas the comparison at the same proportion of maturity reveals developmental difference.

If a structure is neither early nor late maturing and proceeds to its mature weight on the same pattern as liveweight then comparisons on either basis give the same answer.

Exactly the same principle applies to comparisons within a system such as the proportions of individual muscle weight to total muscle weight ('muscle-weight distribution'). Early and late maturing muscles comprise different proportions of total muscle weight at different stages of maturity and hence can be compared in a biologically sensible way only at the same proportion of maturity of the total musculature.

It is clear from the above that certain information is essential if meaningful comparisons are to be made, namely:

(a) The mature weight of the animals and the mature weights of all parts of the animal which are to be studied.
(b) The maturing patterns of the whole animal and of its component parts.

It is apparent that the nutritional history of animals will play a large part in these data and to assess animals for all conditions will be an impossible task. This should not daunt us from first gaining the necessary information for one set of nutritional conditions from which basis further sets of conditions applicable to commercially likely, or biologically interesting, situations can be developed. The simplest baseline is to work with *ad lib* near optimum nutrition for full exploitation of developmental potential.

Assessing development relative to mature liveweight requires a definition which is a question that will not be resolved to everyone's satisfaction. The definition used by Thompson[26] is suggested as a practical basis in *ad lib* fed animals, namely 'An animal is accepted as mature when it has reached at least 85 % of its asymptote for the exponential relationship between body weight and cumulative feed consumed and the average increment in weekly body weight for at least 10 weeks prior to slaughter is not significantly different from zero.

Weight of mature body components can be determined only by dissection of mature animals if complete data are required. Various techniques (e.g. tritiated water) are available, however, for estimation of some aspects of body composition if less complete information is necessary. Maturing patterns are then derived using data from immature and mature animals. Examples of the shape of maturing patterns of the three main carcass tissues relative to liveweight are shown in Fig. 1.

With some tissues, and this particularly applies to some individual muscles, the growth in the early post-natal phase is vastly different from the remainder of post-natal growth and different considerations will be needed in comparison of very young animals. The earlier system used by Lohse *et al.*[27] in recording growth in multiples of weights at birth, is probably the best basis of measurement for pre-weaning comparisons of muscle-weight distribution. Further work is needed to compare the relationship between early post-natal changes and multiples of birth weight on the one hand and

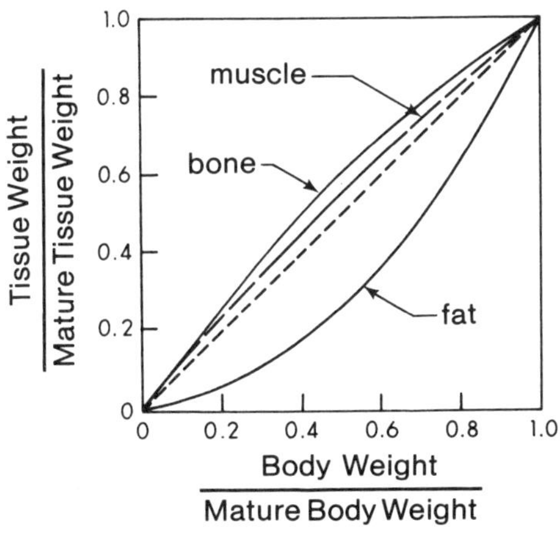

FIG. 1. The progress of muscle, bone and fat to their mature weights in the body, relative to the progress of the total body to its mature weight.

body weight on the other. There may well be a need for different bases of comparison for different tissues in this early post-natal period. Butterfield and Berg[28] showed that there was a closer relationship between total muscle weight and muscle-weight distribution than between age and muscle-weight distribution. However, from about 20 % of mature weight onwards the mature weight at which the growth of each tissue is aimed seems to be the dominant factor for all tissues.

The studies relating to mature size suggest that comparisons of the proportions comprised by structures within genotypes or 'sexes' of different mature weights fall into the following categories.[21]

(i) *Those structures which, in animals of different mature size, should be compared at the same proportion of maturity.* These are structures, the maturity patterns of which are different from that of the whole animal or total tissue. Hence, these structures are constantly changing in their proportion of the whole animal or total tissue. They must have similar maturing patterns in the genotypes or 'sexes' being compared so that the changes associated with stage of maturity are similar. Fortunately a large proportion of body structures will probably fit these criteria. Examples are carcass bone, carcass fat and body organs as proportions of liveweight, muscle-weight distribution, bone-weight distribution and fat-weight distribution.

(ii) *Those which may be compared at equal weight irrespective of mature size.* These are structures, the maturity patterns of which do not differ greatly from that of the whole animal or total tissue and hence retain a near constant proportion of the whole animal or tissue. Examples are total carcass muscle as a proportion of liveweight, fat partitioning between carcass and non-carcass depots as proportions of total fat weight and fat partitioning between the various non-carcass depots as proportions of total fat weight.

(iii) *Those structures which cannot be sensibly compared biologically at any one weight or any one proportion of mature weight.* These are those structures which have different maturing patterns in the genotypes being compared.

Examples of these are fortunately few in the more usual types of comparisons, however, the head weight as a proportion of liveweight may be one such example in comparisons of horned and polled breeds. There may well be many examples in comparing extremely different types of animals.

Many comparisons (among genotypes or sexes) will be made of animals which are of similar mature size and so comparisons of body proportions at equal weight will be the same as comparisons at equal stage of maturity. However, to ensure that comparisons are embracing only developmental differences and not a mixture including some stage of maturity differences, it is necessary that we know the mature weight of the genotypes or sexes and the mature weights of their component structures. If studies are to be made of proportions within systems (e.g. muscle-weight distribution) it will not be sufficient to know that the mature liveweights are the same but rather that mature total muscle weights are the same. We lack this information for our existing genotypes which must act as the markers for 'improved' animals.

There are few if any body composition characteristics of commercial interest which can be compared on an age basis, except the overall size of the animals. Under similar nutritional conditions, weight at a given age is probably the most important characteristic of meat animals. However, the internal proportions of animals seem to be better related to weight and even better to stage of maturity than to age. Butterfield and Berg[28] showed that changes in muscle-weight distribution in young calves were closely related to total muscle weight and poorly related to age under an extreme nutritional regime. However, because of the close relationship between age and weight in *ad lib* fed animals many comparisons on an age basis will give similar results to comparisons at equal weight in animals of similar mature size.

Until more evidence is available, it would seem that the best comparisons of proportional composition in very young animals should be made on a weight basis measured in multiples of birth weight. After the achievement of around 20% of mature weight most comparisons are best made at the same stage of maturity.

4. MEAT PRODUCTION FROM INTACT MALES

Many studies have shown entire males as compared to castrates, to have superior growth rate and feed conversion efficiency and to produce leaner carcasses. A symposium held at The Meat Research Institute in Bristol dealt with production from entire males in all its ramifications.[29] Negative aspects of use of entire males have been reported as aggressive behaviour, undesirable odours and flavours, lower quality grades, less tender meat and undesirable colour of lean.

In spite of the drawbacks, production from entire males is the chosen alternative in many countries. Bull production is the major system in most

Continental European countries but steers predominate in the United Kingdom, the Americas, Australia and New Zealand.

Production from boars is considered to present greater problems than from bulls because of unpleasant boar taint odour. Nevertheless in some countries boar production is the most prevalent system. In Ireland, Hanrahan and O'Grady[30] estimated that 85% of male pigs slaughtered were entire and very few pigs are now castrated in intensive piggeries in Australia. A survey of consumer reaction to boar meat in six European countries concluded that boar meat would be acceptable to most consumers.[31] Mottram *et al.*[32] reported that taste panelists, who were sensitive to boar taint, did not give boar bacon lower scores than that from barrows for flavour and overall acceptability, and in fact found boar bacon to be more tender.

Malmfors and Lundstrom[31] suggested more research is necessary to develop a screening test for carcasses which might produce undesirable levels of boar taint. A promising alternative procedure was reported by Williamson and Patterson[33] who used autoimmunisation selective for 5α-androstenone which resulted in a reduction in androstenone levels in adipose tissue to 1/3 of the level found in untreated boars. The development of immunologic castration[34] may eventually overcome many of the objections both to castration and to non-castration. As pointed out by Short[35] its main potential lies in cattle where the process may be delayed for a sufficient time to allow the exploitation of the growth potential of the entire without the problems which may be encountered later in marketing bulls or in adult castration.

Aggressive behaviour was observed to be no particular problem in young bulls in a feedlot holding approximately 200 bulls.[36] Bulls did, however, show greater agonistic and sexual (mounting) activity than steers. Smaller groups of bulls (less than ten per pen) under more crowded conditions ($7\,m^2$/head) showed less agonistic behaviour than the large (approx. 200) groupings. Regrouping of bulls resulted in marked increase in agonistic and sexual behaviour which returned to normal low levels in a matter of days.[36]

Bulls are more liable to produce dark cutting meat than steers; but Price and Tennessen[37] found dark cutting could be minimised by preslaughter management which avoided mixing during the preslaughter procedures. If mixing was practised, minimising the time interval from mixing to slaughter was beneficial.

It is expected that bull carcasses will have less fat cover and marbling than carcasses from steers. Under the US grading system this could result in a lower quality grade. Seideman *et al.*,[38] reporting US research, found bull beef to be less acceptable than steer beef. Hawrysh *et al.*[39] in Canada found only slight differences in overall acceptability in eating quality between beef from young bulls in comparison with that from steers and found bull beef was in the acceptable range.

Considerable interest is currently being shown in New Zealand and Australia in the possibility of reducing the proportion of male lambs which are castrated. In areas of good nutrition and assured turn-off of prime lambs it seems ridiculous to castrate lambs within a few weeks of slaughter—a procedure which has been carried out with little consideration of its need. Pressure from animal welfare groups may be added to economic factors to result in a reduction of castration of lambs in the future.

Problems associated with production and quality of meat from entire males, although circumvented to some extent, cannot be considered to have been eliminated. Because of the inherent superiority of the entire male in growth, efficiency and lean meat yield, meat production from entire males will undoubtedly continue to increase.

5. MEAT PRODUCTION FROM DOUBLE MUSCLED ANIMALS

Extreme muscular hypertrophy occurs in many species. In cattle it is known as double muscled, cullard, dopellander, and groppa dopia, etc. Double muscling is manifest by a high percentage of muscle, a high muscle:bone ratio and a low proportion of fat. Carcass yield is also high compared to normal animals—a topic reviewed in King and Menissier.[40] Geay *et al.*[41] found double-muscled Charolais bulls had lower feed intake and liveweight gain than normal. Muscle and protein deposition potential were similar while lipid deposition differed markedly. The double-muscled bulls were more efficient in protein deposition relative to energy consumed but were less efficient in energy stored. They described the double-muscled animals as being an extreme type of late maturing animals. In cattle, besides its obvious effect on carcass composition, double-muscling has a number of other effects which affect productivity. Menissier[42] in a review stated that double-muscled calves were heavier at birth, which contributed to calving difficulty. Fertility in females was impaired, viability was reduced and milk production was reduced. Menissier found these negative effects to be quite variable in their severity.

Inheritance of muscling differences is considered to be polygenic but the extreme double muscling condition generally has a major gene effect superimposed on polygenic and environmental variation.[43] The expression of the major gene in homozygous or heterozygous condition depends on the genetic background (e.g. modifier genes). Penetrance of the major gene seems to be complete when homozygous but may be incomplete (reduced or non-existent) in the heterozygote. Lauvergne *et al.*[44] stated that variation in the expression in the heterozygote may be explained in terms of the evolution of dominance as developed by Fisher.[45] When the character is considered advantageous, and is the object of favourable selection, it tends to become dominant, as in the Piemontese (i.e. the heterozygote is expressed). When it is considered undesirable, as in dairy breeds, it tends to become recessive (i.e. little expression of the heterozygote).

Ollivier[46] presents evidence for a single major gene being the main determinant of muscular hypertrophy in the Belgian Pietrain breed of pigs.

Muscling characterised by percentage muscle in a carcass, or by muscle:bone ratio, shows a great deal of variation in different breed types. Selection for 'dairy character' in North America has resulted in a low percentage of muscle and low muscle:bone ratios. Early fattening British breeds have moderate muscling and Continental breeds, selected for meat characteristics, have high muscle percentage and muscle:bone ratios. Berg *et al.*,[11] in a report involving male progeny of eight sire breeds used on Danish dairy breeds, found a variation in muscle:bone ratios of 4·1:1 for Chianina crossbreds to 4·9:1 for Limousin crossbreds. Berg and Shahin[47] compared young double muscled bulls (described as moderate to extreme) with Hereford and a synthetic strain combining Charolais, Angus and Galloway breeds. They reported muscle:bone ratios of 5·2:1 for the Herefords, 5·6:1 for the synthetics and 6·5:1 for the double-muscled groups.

Heavy muscling seems also to influence yield of muscle relative to liveweight. Callow[48] suggested that cattle (British beef breeds) had about 1/3 of their liveweight as muscle tissue. In the breed crosses reported by Berg *et al.*[11] muscle as a percentage of liveweight varied from 35 % (for the Hereford) to 41 % (for the Blonde D'Aquitaine crossbreds). In the three groups of Berg and Shahin[47] muscle relative to liveweight varied from 33 % for Herefords and 39 % for synthetics to 41 % for the double-muscled group. In the most extreme double-muscled animals of Berg and Shahin[47] the muscle:bone ratio exceeded 8:1 and muscle represented 50 % of liveweight.

Double-muscled animals seem to have differential hypertrophy of muscles in different regions of the carcass as well as differential hypotrophy of some bones. Shahin and Berg[16] reported that, at the same total side muscle weight, double-muscled animals had a significantly lower percentage of muscle in the distal limb region and in muscles of the neck and thorax than did normal breed groups. Double-muscled animals exceeded the other groups in percentage of muscle in the hip and stifle region.

It seems probable that, to a large extent, the differential hypertrophy, as measured by the proportionate muscle:weight distribution, is due to the variation in the proportion of connective tissue in each muscle and hence in the proportion of muscle fibres—as proposed by Berg and Butterfield[9] on the basis of the study of Hanset and Ansay.[49] More recently Boccard[50] produced valuable information on the relative proportions of collagen in a range of muscles of normal and double-muscled Charolais cattle and introduced the concept of the differential dilution of collagen by the hypertrophy of the muscle.

At the same side bone weight, double-muscled animals had lighter femurs than the other groups.[16]

Shahin and Berg[16] compared muscle:bone ratios in various regions between double-muscled, Hereford and a synthetic strain of cattle. These are shown in Table 1. Muscle:bone superiority was most pronounced in the

TABLE 1
MUSCLE:BONE RATIOS AND HYPERTROPHY RATIOS BY BREED GROUP FOR DIFFERENT ANATOMICAL REGIONS

Anatomical region	*Breed*				
	Muscle:bone ratio[a]			*Hypertrophy ratio*[b]	
	HE	*SY*	*DM*	*DM/HE*	*DM/SY*
Proximal pelvic limb	6·31	6·77	8·39	1·33	1·24
Distal pelvic limb	3·09	3·10	3·15	1·02	1·02
Back and loin	4·58	4·98	5·58	1·22	1·12
Proximal thoracic limb	4·50	4·66	5·35	1·19	1·15
Distal thoracic limb	1·88	1·96	2·12	1·13	1·08
Total side	5·2	5·6	6·5	1·25	1·16

[a] Relevant muscle and bone groups adjusted to equal total side muscle weight.
[b] Muscle:bone ratio of DM divided by muscle:bone ratio of HE and SY, respectively.
HE—Hereford, SY—synthetic, DM—double muscled.

proximal pelvic limb, followed by the back and loin and the proximal thoracic limb, being less pronounced in the distal limbs.

Several approaches have been taken in different countries to exploit the double-muscling condition for beef production. In the Piemontese breed[51] and in the Belgian Blue breed[52] selection is within the purebred and reduction of deleterious side effects is accomplished by genetically modifying the pleiotropic effects of the double-muscled gene. In the Belgian Blue, however, Caesarean births are accepted as normal. In France an attempt has been made to develop a specialised double-muscled sire line to be used in a terminal crossbreeding system.[53] Another alternative is to attempt to increase muscling in normal cattle by selection for polygenic differences.

Liboriussen[54] compared progeny of double-muscled Belgian Blue and Piemontese bulls from Danish Red and Danish Friesian cows. Crossbred young bulls from double-muscled sire breeds were more muscular than those from more normal sire breeds. Calves sired by Belgian Blue bulls had the same amount of calving difficulty as those from Charolais sires, although the former were 3 kg lighter. Piemontese sired calves, which were 5 kg lighter than those from Charolais, had about the same incidence of calving difficulty. This indicates that, at equal birth weights, double-muscled sired calves are likely to present more calving difficulty. Liboriussen[54] reported that meat quality of the young bulls sired by Belgian Blue and Piemontese was equal to those from Charolais and in general not much different than the quality from other sire breeds used.

The French development of specialised double-muscled sire lines for terminal crossbreeding was not developed to the extent whereby its efficiency could be assessed.[53]

Double-muscling results in a greater yield of muscle to liveweight, a greater yield of muscle in a carcass, some improvement in the muscle-weight distribution and for some markets an improvement in 'quality'. In cattle, extreme double-muscling is beset with many biological problems—increased calving difficulty, increased early mortality, reduced reproductive performance and a shortened productive life span. Economic viability of double-muscled enterprises is very local and depends on enhanced value of the product because of either yield or quality offsetting the increased costs of production resulting from biological shortcomings. In these respects a double-muscled enterprise is similar to any other meat production enterprise—it is another example of economic exploitation by man of a unique type within domestic species. Obviously such enterprises must be profitable to be viable.

6. POSSIBLE INFLUENCES OF SELECTION PRACTICES ON CARCASS COMPOSITION

Selection in most meat animal species emphasises adaptation and reproduction, growth rate and efficiency and carcass composition. We have seen earlier that extreme muscling, at least as manifest by the double-muscled condition, has a negative influence on female reproductive performance in cattle. This along with the general low fertility and long generation interval in cattle, has led Vissac *et al.*[55] to suggest that within breed selection is not a viable proposition. They say it is more efficient to exploit breed differences in traits of economic importance by cross-breeding. Complementarity between the muscle growth potential of large-sized breeds and the maternal breeding value of hardy (adapted) breeds can be exploited. Such a system implies different selection criteria for paternal and maternal strains to be used in crossbreeding.

Major emphasis in beef cattle selection has recently been for more rapid growth rate. Andersen *et al.*[56] reported that Shorthorns selected for high yearling weight were less fat at a constant weight than were controls but there was little difference in fatness at the same age when selected animals were larger. Cundiff[57] reported that steers from breed groups representing biological types of larger size had higher percentages of retail product and a lower percentage of fat trim than those representing types of smaller size, which are manifestations of the influence of mature size and the later development of fat in the larger type breeds relative to liveweight.

McCarthy[58] concluded from mouse studies that when selection for growth rate was on *ad lib* feeding the most obvious correlated response was an altered rate of food consumption. However, food consumed per gram liveweight at a constant weight did not seem to have been altered. There was also some screening for genotypes that utilise energy more efficiently. Selection for high growth rate also caused changes in the pattern of adipose tissue deposition resulting in leaner carcasses. The effect was greater when exercised at older ages (i.e. when more fat is normally deposited). Selection on a fixed level of feeding resulted in a modest response in gain and an increase in efficiency and probably less fat deposited.

Robertson[59] explained the results from selection for early growth rate as a response to appetite. Since in the early stages of growth, relatively more lean is deposited, increased appetite would result in greater (lean) growth. Later the increased appetite leads to surplus energy for fattening and therefore to fatter animals at maturity. Selection at later ages would presumably distinguish between animals which have a predisposition to lean rather than fat growth.

If the general model holds for larger meat animal species, then selection for high or low body weight at an early age should have little effect on the rate of fat deposition relative to body weight, and consequently animals from high and low body weight strains should have a similar percentage fat in the body at the same body weight, although the high body weight strain should be fatter at maturity. In animals selected for high body weight at a later age, fat should be later maturing and therefore these animals would be leaner at given weights—but probably not too different at maturity to those from strains selected for low body weight.

Studies by Thompson *et al.*[24,25] provide results for assessing the influence of nutritional effects of milk on pre-weaning selection for growth rate in lambs. They examined the effect of selection for high and low weaning weight on body composition in sheep and showed that fat, in the high weaning weight selection strain, was earlier maturing than in the low weaning strain. Furthermore, animals from both the high and low weaning weight selection strains had a similar percentage of fat in the body at maturity even though they had widely different mature weights.

It was suggested by Thompson *et al.*[24,25] that the earlier maturing pattern for fat in the high weaning weight strain was due to an increased intake of milk during pre-weaning growth. Pattie[60] showed that ewes from the high weaning weight strain had greater milk production than ewes from the low weaning weight strain, although (perhaps as expected) the differences were proportional to the differences in the ewe body weight.[61] However in the pre-weaned lamb an increase in body fat is consistent with an increase in milk intake in the lamb.[62] As animals from the high and low weaning weight strains all progressed to the same percentage fat at maturity, it follows that fat would be relatively earlier maturing in the 'weight plus' and later maturing in the 'weight minus' animals relative to the randoms. However the earlier maturing of fat in the weight plus group is not a result of the increased appetite of the lambs (as in McCarthy's model) but in increased energy intake because of the milk production of their dams. After the lambs of each group were weaned, they proceeded to a common mature composition which in fact had not been affected by the selection practised.

From these results it is interesting to speculate as to the possible effect of weight selection at later ages on the changes in body composition of the animal. In the development of breeds or strains of different mature size, it is likely that selection for different body size occurred at about the time replacement animals were selected to enter the breeding flock. Given this, strain and breed differences in compositional characteristics should be consistent with the effect of selection for weight at later ages. In a comparison of large and small strains of Merino rams, Butterfield *et al.*[18–20] showed that both strains had similar maturity coefficients for fat and, at maturity, had about the same proportion of fat in the body. Therefore it is proposed that the differences in body composition in strains of sheep selected for body weight at different post-weaning ages, would be largely a function of their differences in mature size. However, as the heritabilities for body weight may vary post-weaning, obviously the response in body weight, and therefore in mature size, would also vary with different selection ages. Therefore selection for body weight at a number of different ages may still result in different body compositions at the same body weight. These differences, however, may simply reflect differences in stage of maturity rather than any intrinsic differences between the strains in their maturing patterns of fat in the body.

Selection in cattle for high growth rate under tropical conditions has different components than those found in temperate conditions. Frisch[5] showed that selection for high growth rate in a stressful environment may be a result of improvement in resistance to the stressors (disease, parasites or high temperature). Conversely, selection for growth rate in a low stress (temperate) environment may actually decrease growth response in a high stress environment because resistance to environmental stressors may actually decrease. This thesis was supported by results that Brahman and Brahman cross cattle grew better than Bos taurus types under tropical conditions but this situation was reversed under temperature conditions or when the stressors were removed artificially.

Frisch[5] found similar results for fertility where Brahman cross cattle were superior to Bos taurus types in the tropics but the latter showed a much greater favourable response under temperate conditions which resulted in superior reproductive performance compared to the Brahman crossbreds.

Survival under dry tropical conditions depends to some extent on the ability to survive under reduced feed availability. On the other hand, selection for performance (growth rate) in temperate regions has favoured animals with larger appetites which would provide no advantage for survival in the tropics.

Selection for adaptability under tropical conditions, as illustrated by adaptability of Zebu cattle and their crosses in many tropical countries, has different components than selection for adaptability under temperate conditions and cold winters. Increased appetite and increased storage of fat is compatible with cold winter survival, but may have a negative effect on survival in the tropics.

The possible influences which selection for adaptability may have on carcass composition have not been thoroughly studied. One might surmise that lower appetite and lower growth rate could result in a lower proportion of fat under tropical conditions and that a higher appetite and faster growth rate could lead to a higher proportion of fat under temperate conditions.

Little information is available as to the correlated responses expected when selection is for body composition *per se*. Selection for heavier muscling in the Belgian Pietrain breed of pigs resulted in increased halothane sensitivity and possibly a lower fertility in boars.[63] Evidence that the more extreme double-muscled cattle are less well adapted is overwhelming;[40] but no research is available which compares animals with heavier muscling as a result of polygenic variation in the absence of the major double-muscled gene. Until such results are available it would seem wise to avoid extremes in selection for muscular hypertrophy.

7. REFERENCES

1. Tennessen, T. and Hudson, R. J. (1981). *Appl. Anim. Ethol.*, **7**, 87.
2. Lawrie, R. A. (1974). *Meat science*, Pergamon Press, Oxford.
3. A.M.R.C. Report (1983). Aust. Meat Res. Committee, Sydney.
4. Hudson, R. J. (1976). *Potential for meat production from marginal land resources*, Misc. Publ., Department of Animal Science, University of Alberta, Edmonton, Alberta.
5. Frisch, J. E. (1984). In: *Beef cattle science handbook* (Baker, F. H. and Milles, M. E. (Eds)) Winrock International, Arkansas.
6. Spedding, C. R. W. (1983). In: *Proc. int. symp. on beef prod.*, Kyoto, Japan, p. 1.
7. Fowler, V. R., Bichard, M. and Pease, A. (1976). *Anim. Prod.*, **23**, 365.
8. Berg, R. T. and Walters, L. E. (1983). *J. Anim. Sci.*, **57** (Suppl. 2), 133.
9. Berg, R. T. and Butterfield, R. M. (1966). *Anim. Prod.*, **8**, 1.
10. Berg, R. T. and Butterfield, R. M. (1976). *New concepts of cattle growth*, Sydney University Press, Sydney.
11. Berg, R. T., Andersen, B. B. and Liboriussen, T. (1978). *Anim. Prod.*, **26**, 245.
12. Butterfield, R. M. (1965). *Proc. New Zealand Soc. Anim. Prod.*, **25**, 152.
13. Bergstrom, P. L. (1978). In: *Patterns of growth and development in cattle. Current topics in veterinary medicine, Vol. 2* (de Boer, H. and Martin, J. (Eds)), Martinus Nijhoff, The Hague, p. 91.

14. TAYLOR, St. C. S., MASON, M. A. and MCCLELLAND, T. H. (1980). *Anim. Prod.*, **30**, 125.
15. DAVIES, A. S. (1974). *Anim. Prod.*, **19**, 337.
16. SHAHIN, K. A. and BERG, R. T. (1983). *62nd Ann. Feeder's Day Rept.*, University of Alberta, Edmonton, Alberta, p. 34.
17. TAYLOR, St. C. S. (1968). In: *Growth and development of mammals* (Lodge, G. A. and Lamming, G. E. (Eds)), Butterworths, London, p. 267.
18. BUTTERFIELD, R. M., GRIFFITHS, D. A., THOMPSON, J. M., ZAMORA, J. and JAMES, A. M. (1983). *Anim. Prod.*, **36**, 29.
19. BUTTERFIELD, R. M., ZAMORA, J., JAMES, A. M., THOMPSON, J. M. and REDDACLIFF, K. J. (1983). *Anim. Prod.*, **36**, 461.
20. BUTTERFIELD, R. M., ZAMORA, J., JAMES, A. M., THOMPSON, J. M. and WILLIAMS, J. (1983). *Anim. Prod.*, **36**, 165.
21. BUTTERFIELD, R. M. and THOMPSON, J. M. (1983). *Anim. Prod.*, **37**, 423.
22. BUTTERFIELD, R. M., REDDACLIFF, K. J., THOMPSON, J. M., ZAMORA, J. and WILLIAMS, J. (1984). *Anim. Prod.*, **39**, 259.
23. BUTTERFIELD, R. M., ZAMORA, J., THOMPSON, J. M., REDDACLIFF, K. J. and GRIFFITHS, D. A. (1984). *Anim. Prod.*, **39**, 251.
24. THOMPSON, J. M., BUTTERFIELD, R. M. and PERRY, D. (1985). *Anim. Prod.*, **40**, 71.
25. THOMPSON, J. M., PARKS, J. R. and PERRY, D. (1985). *Anim. Prod.*, **40**, 55.
26. THOMPSON, J. M. (1983). Feed intake, growth and body composition in flocks of Merino sheep selected for high and low weaning weight, Ph.D. Thesis, University of Sydney, Sydney, Australia.
27. LOHSE, C. L., MOSS, F. P. and BUTTERFIELD, R. M. (1971). *Anim. Prod.*, **13**, 117.
28. BUTTERFIELD, R. M. and BERG, R. T. (1966). *Proc. Aust. Soc. Anim. Prod.*, **6**, 298.
29. RHODES, D. N. (1969). *Production from entire male animals*, J. & A. Churchill Ltd, London.
30. HANRAHAN, T. J. and O'GRADY, J. F. (1982). *Proc. Annu. Meeting. E.A.A.P. Leningrad, 1982.*
31. MALMFORS, B. and LUNDSTROM, K. (1982). *Proc. Annu. Meeting E.A.A.P. Leningrad, 1982.*
32. MOTTRAM, D. S., WOOD, J. D. and PATTERSON, R. L. S. (1982). *Anim. Prod.*, **35**, 75.
33. WILLIAMSON, E. D. and PATTERSON, R. L. S. (1982). *Anim. Prod.*, **35**, 353.
34. ROBERTSON, I. S., WILSON, J. C. and FRASER, H. M. (1979). *Vet. Rec.*, **105**, 556.
35. SHORT, R. V. (1980). In: *Growth in animals* (Lawrence, T. L. J. (Ed.)), Butterworths, London, p. 25.
36. TENNESSEN, T. (1983). An ethological investigation into the husbandry of bulls, Ph.D. Thesis, University of Alberta, Edmonton, Alberta.
37. PRICE, M. A. and TENNESSEN, T. (1981). *Can. J. Anim. Sci.*, **61**, 205.
38. SEIDEMAN, S. C., CROSS, H. R., OLTJEN, R. R. and SCHANBACHER, B. D. (1982). *J. Anim. Sci.*, **55**, 826.
39. HAWRYSH, Z. J., BERG, R. T. and PRICE, M. A. (1979). *Can. Inst. Food Sci. Technol. J.*, **12**, 72.
40. KING, J. W. B. and MENISSIER, F. (Eds) (1982). *Muscle hypertrophy of genetic origin and its use to improve beef production*, Martinus Nijhoff, The Hague.

41. GEAY, Y., ROBELIN, J., VERMOREL, M. and BERANGER, C. (1982). In: *Muscle Hypertrophy of Genetic Origin and its Use to Improve Beef Production* (King, J. W. B. and Menissier, F. (Eds)), Martinus Nijhoff, The Hague, p. 74.
42. MENISSIER, F. (1982). In: *Muscle hypertrophy of genetic origin and its use to improve beef production* (King, J. W. B. and Menissier, F. (Eds)), Martinus Nijhoff, The Hague, p. 23.
43. MENISSIER, F. (1982). In: *Muscle hypertrophy of genetic origin and its use to improve beef production* (King, J. W. B. and Menissier, F. (Eds)), Martinus Nijhoff, The Hague, p. 387.
44. LAUVERGNE, J. J., VISSAC, B. and PERRAMON, A. (1963). *Ann. Zootech.*, **12**, 133.
45. FISHER, R. A. (1930). In: *The genetic theory of natural selection*, Oxford University Press, Oxford, p. 48.
46. OLLIVIER, L. (1982). In: *Muscle hypertrophy of genetic origin and its use to improve beef production* (King, J. W. B. and Menissier, F. (Eds)), Martinus Nijhoff, The Hague, p. 429.
47. BERG, R. T. and SHAHIN, K. A. (1983). *62nd Ann. Feeders' Day Rept.*, University of Alberta, Edmonton, Alberta, p. 32.
48. CALLOW, E. H. (1948). *J. Agric. Sci., Camb.*, **38**, 174.
49. HANSET, R. and ANSAY, M. (1972). *Ann. Med. Vet.*, **116**, 17.
50. BOCCARD, R. (1981). In: *Developments in meat science—2* (Lawrie, R. A. (Ed.)), Applied Science Publishers Ltd, London, p. 1.
51. MASOERO, G. and POUJARDIEU, B. (1982). In: *Muscle hypertrophy of genetic origin and its use to improve beef production* (King, J. W. B. and Menissier, F. (Eds)), Martinus Nijhoff, The Hague, p. 450.
52. HANSET, R. (1982). In: *Muscle hypertrophy of genetic origin and its use to improve beef production* (King, J. W. B. and Menissier, F. (Eds)), Martinus Nijhoff, The Hague, p. 437.
53. MENISSIER, F. (1982). In: *Muscle hypertrophy of genetic origin and its use to improve beef production* (King, J. W. B. and Menissier, F. (Eds)), Martinus Nijhoff, The Hague, p. 480.
54. LIBORIUSSEN, T. (1982). In: *Muscle hypertrophy of genetic origin and its use to improve beef production* (King, J. W. B. and Menissier, F. (Eds)), Martinus Nijhoff, The Hague, p. 637.
55. VISSAC, B., FOULLEY, J. L. and MENISSIER, F. (1982). *Proc. world congress sheep and beef cattle breeding, Vol. 1: Technical* (Barton, R. A. and Smith, W. C. (Eds)), Dunmore Press, Palmerston North, New Zealand, p. 101.
56. ANDERSEN, B. B., FREDEEN, H. T. and WEISS, G. M. (1974). *Can. J. Anim. Sci.*, **54**, 117.
57. CUNDIFF, L. V. (1982). In: *Proc. world congress sheep and beef cattle breeding, Vol. 1: Technical* (Barton, R. A. and Smith, W. C. (Eds)). Dunmore Press, Palmerston North, New Zealand, p. 71.
58. MCCARTHY, J. C. (1982). *Proc. world congress sheep and beef cattle breeding, Vol. 1: Technical* (Barton, R. A. and Smith, W. C. (Eds)), Dunmore Press, Palmerston North, New Zealand, p. 415.
59. ROBERTSON, A. (1982). *Proc. world congress sheep and beef cattle breeding, Vol. 1: Technical* (Barton, R. A. and Smith, W. C. (Eds)), Dunmore Press, Palmerston North, New Zealand, p. 427.

60. Pattie, W. A. (1965). *Aust. J. Exp. Agric. Anim. Husbandry*, **5**, 361.
61. Taylor, St. C. S. (1973). *Proc. Brit. Soc. Anim. Prod.*, **2**, 15.
62. Black, J. L. (1974). *Aust. Soc. Anim. Prod.*, **10**, 211.
63. Sellier, P. (1982). In: *Muscle hypertrophy of genetic origin and its use to improve beef production* (King, J. W. B. and Menissier, F. (Eds)), Martinus Nijhoff, The Hague, p. 537.

Chapter 2

MYOFIBRILLAR CYTOSKELETAL PROTEINS OF VERTEBRATE STRIATED MUSCLE

KOSCAK MARUYAMA

Department of Biology, Chiba University, Chiba 260, Japan

1. INTRODUCTION

The fundamental role of the actin-myosin-ATP interactions in muscle function has been well established[1] and the universal occurrence of the contractile proteins in cells has been demonstrated.[2] In addition to the contractile proteins, myosin and actin, attention has been focussed on the regulatory proteins, troponin and tropomyosin in striated muscle in relation to the control of the contraction and relaxation through the level of free calcium ions.[3]

Recently, the physiological importance of the cytoskeletal structures, microtubules, intermediate filaments and microfilaments (actin) has been emphasised in the understanding of various cell functions, including cell motility and dynamic changes in cell shape.[4] In striated muscle, myosin and actin filaments, together with Z lines, are regarded as the main cytoskeletal structures, because both filaments essentially form myofibrillar structures. In addition, there are known to be other elements of cytoskeletal framework in myofibrils which are involved in maintaining shape and function. However, investigation of the latter has been limited in comparison with the remarkable progress made in the studies on contractile and regulatory systems. For example, the presence of desmin (skeletin) as intermediate filaments running longitudinally through neighbouring Z lines has been shown only recently. The present review aims at documenting the present state of knowledge on the less well understood aspects of myofibrillar cytoskeletal proteins in the hope of stimulating

TABLE 1
MYOFIBRILLAR PROTEINS OF VERTEBRATE SKELETAL MUSCLE

Proteins	*MW*	*Content* % (*w*/*w*)	*Localisation*	*Discovery*
Contractile proteins				
Myosin	500 kDa	43	A band	Hungary (1942)
Actin	43 kDa	22	I band	Hungary (1942)
Regulatory proteins				
MAJOR				
Troponin (C, I, T)	70 kDa	5	I band	Japan (1965)
Tropomyosin	(33 kDa) × 2	5	I band	UK (1948)
MINOR				
M protein	165 kDa	2	A band	UK (1972)
C protein	135 kDa	2	A band	UK (1973)
H protein	69 kDa	< 1	A band	UK (1983)
F protein	121 kDa	< 1	A band	UK (1973)
I protein	50 kDa	< 1	A band	Japan (1977)
X protein	152 kDa	< 1	A band	UK (1983)
β-actinin	35 kDa + 32 kDa	< 1	End of actin filament	Japan (1965)
γ-actinin	35 kDa	< 1	I band (?)	Japan (1980)
Cytoskeletal proteins				
INTERMEDIATE FILAMENT				
Desmin (skeletin)	53 kDa	1	Periphery of myofibrils and Inter Z lines	Denmark (1977) USA (1977)
Vimentin	55 kDa	1		FRG (1978) USA (1978)
ELASTIC FILAMENT				
Connectin (titin)	2×10^6 kDa	10	A–I junction	Japan (1977) USA (1979)
Z DISC CONSTITUENT				
α-actinin	(95 kDa) × 2	2	Z disc	Japan (1965)
Z-protein	50 kDa	< 1	Z disc	Japan (1979)
Eu-actinin	43 kDa	< 1	Z disc	Japan (1981)
34 K protein	34 kDa	< 1	Z disc	Japan (1982)
Z-nin	400 kDa	< 1	Z disc	Japan (1981)
ABP (filamin)	(250 kDa) × 2	< 1	Z disc	USA (1976)
OTHER PROTEINS				
Vinculin	130 kDa	< 1	sarcolemma	USA (1979)
Nebulin	800 kDa	3	I band	USA (1981)
Paratropomyosin	34 kDa	< 1	A–I junction	Japan (1983)

Compiled mainly from Obinata *et al.*[5] and Yates and Greaser.[6]

further investigation. Thus, for example, vinculin, a protein which attaches actin to the cell membrane, is described in some detail.

In Table 1, the myofibrillar structural proteins of vertebrate skeletal muscle are classified into three categories: contractile, regulatory, and cytoskeletal. It is difficult to separate the latter two groups clearly. Some of the minor regulatory proteins (e.g. M protein, C protein, and β-actinin) are regarded as regulating structure. α-Actinin and eu-actinin (localised in Z lines) may be included in this category. In the present review, proteins of the Z-disc are listed under cytoskeletal proteins. The minor proteins associated with actin and myosin filaments in the sarcomere will be excluded from consideration.[5]

2. CELL MEMBRANE ATTACHMENT PROTEINS

2.1. Spectrin

It is well established that spectrin and related proteins (fodrin, calspectin, TW260/240, etc.) play an essential role in maintaining the shape and plasticity of cell membranes, linking actin filaments to membrane proteins.[7–10] Although there has not yet been any report on their presence in skeletal muscle, spectrin has been detected in embryonic chicken cardiac myocytes by indirect immunofluorescence technique.[8]

2.2. Vinculin

Location of Vinculin in Non-muscle Cells

Vinculin was discovered by Geiger in 1979[11] as a contaminant during the preparation of α-actinin from chicken gizzard. However, observations by indirect immunofluorescence showed an unexpected distribution: it is localised predominantly in intracellular plaques at the bottom of cultured embryonic chicken cells (skin, heart and gizzard). Vinculin-rich areas are found at the termini of actin-containing fibre bundles. Microinjection of fluorescent, dye-conjugated, vinculin into living fibroblasts revealed that it is distributed as patches at the ends of actin filaments on the lower surface of the cell and also presents a fibrillar appearance beneath the upper surface of the cells.[12] Immunoelectron microscope studies showed that vinculin is confined to the membrane-associated dense plaques.[13] Furthermore, extracellular fibronectin filaments are attached to the vinculin-sites to which actin filaments are linked—suggesting a transmembrane relationship between fibronectin and vinculin.[14]

Preparation of Vinculin

Feramisco and Burridge developed a rapid reproducible procedure of purification of vinculin together with α-actinin and ABP (filamin) from chicken gizzard.[15] Suspensions of gizzard well washed with water are extracted for 30 min at 37 °C in 2 mM Tris, pH 9·0, 1 mM EGTA and 0·5 mM PMSF. The supernatant, adjusted to neutral pH, is added together with 10 mM $MgCl_2$. The precipitate formed by $MgCl_2$ (containing actin) is removed. Ammonium sulphate fractionation is made, to collect proteins precipitatable between 14·9 g and 20·5 g per 100 ml. The main components—vinculin, ABP (filamin) and α-actinin—are separated by DEAE cellulose chromatography. This procedure can be effectively applied to skeletal muscle.[16] The yields were approximately 2, 4, and 6 mg from 100 g of breast, ALD (slow skeletal muscle) and cardiac muscles of chicken, respectively.[16]

Physicochemical Properties of Vinculin

Using SDS electrophoresis or gel filtration, the molecular weight of vinculin is estimated to be 130 kDa, both in denatured and native states.[11] The molecular weight of vinculin is the same irrespective of its origin.[11,15] Vinculin appears as a globular protein with a dimer of 8·5 nm in negatively stained samples.[17]

Several isoforms of vinculin are distinguishable by an electric focussing technique.[16,18] In chicken gizzard vinculin, five isoforms ($\alpha, \alpha', \beta, \beta', \gamma$), ranging in pI from 6·8 to 7·3, are present. α-Vinculin is the most acidic. They have the same molecular weight (130 kDa). Pig heart vinculin also shows microheterogeneity, there being three isoforms, ranging in pI from 6·4 to 6·8. Peptide map analyses of these vinculin isoforms from two different tissues show a remarkable similarity. According to Endo and Masaki,[16] apart from the absolute values of pI, the main bands (α, β, in Geiger's terminology[18]) are common to breast, slow heart and brain of chicken.

Judging from the crossreactivity of antibodies against vinculin with vinculins from different origin, and the similar peptide maps of isoforms, vinculin is thought to be conservative protein. It is desirable to determine the amino acid sequence of vinculin and to clarify the microheterogeneity.

Interaction of Vinculin with Actin

The interaction of vinculin with actin filaments was first studied by Jockusch and Isenberg[19] in comparison with α-actinin that crosslinks actin filaments to form bundles. In fact, vinculin caused the formation of paracrystal-like bundles of actin filaments; but there was no sign of gelation of the actin filaments. Using viscosity measurements, both at high and low shearing forces, it was shown that vinculin lowered the viscosity of polymerising actin solutions. α-Actinin greatly elevates the viscosity of an F-actin solution at low temperatures (5–15 °C), whereas vinculin decreases it, independent of temperature.

Jockusch and Isenberg[19] argue that vinculin does not bind to the ends of actin filaments because, in its presence, elongations of both ends of myosin S1-decorated actin filaments are unaffected. On the other hand, Wilkins and Lin[20] showed that vinculin inhibits the elongation of added actin nuclei under conditions where spontaneous nucleation does not take place. Moreover, the binding of one vinculin molecule to 1500–2000 actin monomers was shown using ^{3}H-labelled vinculin. Thus it was claimed that vinculin binds to the ends (very probably 'barbed end') of actin filaments.

In a detailed investigation, Jockusch and her associates[17] have clearly shown that vinculin does not bind to the ends of actin filaments. They added G-actin to vinculin-coated polystyrene beads, expecting the formation of actin filaments attached by one end to the surface of the beads. Instead, actin filaments were found to be laterally associated with the surface of the beads. The inhibition by vinculin of the elongation of preformed nuclei in a G-actin solution was not observed, in disagreement with the work of Wilkins and Lin.[20] Ferritin-coupled anti-vinculin antibodies bound laterally to the actin filaments in the bundle. Hence it is suggested that vinculin links actin filaments together to interact with the cell membrane.[17]

Vinculin in Skeletal Muscle

Unexpectedly, vinculin has been observed to be located in register at the cytoplasmic surface of muscle fibres of chicken skeletal muscle.[21] Vinculin bands encircle the muscle cell and occur periodically along its length at each sarcomere. The vinculin lattice ('costamere') is divided into two rows which flank the Z line and overlie the I band of the underlying sarcomere. The costameres change their widths in parallel with the changes in the widths of the underlying I bands in stretched and contracted muscles. Therefore, it appears that the two-dimensional lattice consisting of a transverse vinculin element (costamere) and of a longitudinal element (not identified), firmly attach to the myofibrils at the sarcolemma. Some actin filaments at the periphery of myofibrils may attach to the costamere.

Craig and her associates have confirmed (in avian and bovine cardiac muscle) that vinculin is organised at the sarcolemma in a striking array of rib-like bands (costamere).[22] Furthermore, in bovine cardiac muscle, vinculin is present at tubular invagination of the plasma membranes, suggesting that vinculin is related to an extensive system of lateral attachment of myofibrils to the cell membrane.

It would be of much interest to identify the longitudinal element of costameres and the filamentous protein linking Z lines to the costamere. To achieve this, a newly developed technique detecting vinculin-binding proteins may be of great help.[23] In brief, proteins are subjected to SDS-gel electrophoresis, and after renaturation, the gels are treated with ^{125}I-vinculin. Otto[23] has found a 220 kDa vinculin-binding protein in chick embryo fibroblasts and HeLa cells, in addition to actin and vinculin itself. (Is this 220 kDa protein related to spectrin subunits?)

3. INTERMEDIATE FILAMENT

There are three major classes of cytoplasmic filamentous structures in cells: microfilament (actin), intermediate filament and microtubule. These filaments are mainly involved in maintaining cell shape and in cell motility. Since the introduction of the indirect immunofluorescence technique by Lazarides and Weber[24] in 1974, the intracellular localisation of the filaments has been easily visualised.

The diameter of intermediate filaments (7–11 nm) falls between that of thin actin filaments (6–7 nm) and that of thick myosin filaments (15 nm).[25] Intermediate filaments are grouped into five categories: cytokeratin, desmin, vimentin, neural and glial filaments.[26,27] In this section, attention will be directed principally to desmin and vimentin filaments which are present in vertebrate striated muscle.

3.1. Desmin (Skeletin)

Thanks to the pioneering work of Cooke,[28] intermediate filaments have been shown to serve as a shape-determining cytoskeletal network in chicken gizzard and tanenia coli smooth muscles. The constituent protein was estimated[28] to have a molecular weight of 55 kDa. The purified intermediate filament protein from chicken gizzard was termed skeletin by Small and Sobieszek.[29] On the other hand, Lazarides and Hubbard prepared antibodies against the gizzard 50 kDa protein (isolated by a preparative SDS-gel electrophoresis method) and showed its localisation not only in the Z lines but also in the connection between the Z lines of adjacent myofibrils of chicken skeletal muscle.[30] Lazarides termed the protein desmin[30] (from the Greek, 'desmos' = link, bond).

Preparation of Desmin

Desmin is abundantly present in chicken gizzard. Usually, gizzard is used for the isolation of desmin.[31] It is insoluble in neutral salt solution and soluble at extreme pHs.[28,31] Well-washed gizzard residues are extracted with 1 N acetic acid, and the extract is neutralised to produce filamentous sediments. Further purification is achieved by column chromatography in 6 M urea using DEAE-cellulose or hydroxyapatite and DEAE-Sepharose CL-6B columns.[32,33] Desmin has been prepared from mammalian (porcine) skeletal muscle: 24 mg being obtained from 600 g of muscle, representing a yield of 14% of the original content. Thus, in skeletal muscle, the content of desmin is as low as 2·7 mg per 100 g of muscle,[33] in contrast to 750 mg per 100 g in turkey gizzard.[32]

Properties of Desmin

The molecular weight of desmin is estimated to be 55 kDa by SDS-gel electrophoresis[28,29,32,34] and also by sedimentation equilibrium in the presence of 5 M guanidine-HCl.[29] Desmin remains soluble in 10 mM Tris-acetate, pH 8·5: its sedimentation coefficient ($S^{\circ}_{20,w}$) is 5·2 S.[32] It is likely that desmin exists as tetramer at this low ionic strength. The UV absorption spectrum shows a maximum at 278 nm and E_{278} of 5·57. The low value of the extinction coefficient is due to the low content of aromatic amino acids (Table 2). The α helix content is estimated to be about 45% judging from UV circular dichroism measurements.[32]

Steinert and his associates first pointed out that intermediate filaments (baby hamster kidney cell) and keratin filaments have similar structures.[35] The molecule was regarded as consisting of three subunit chains, 48 nm long. Each middle domain has two regions of α helix of 18 nm long interspaced by a nonhelical region (4 nm) at the centre. In each subunit there is a main middle domain, 40 nm in length with head and tail portions. These subunit structures of desmin have been modified by Geisler and Weber who determined the complete amino acid sequence.[36,37] Chicken desmin subunit consists of 463 residues and has a molecular weight of 53 000. The subunit consists of three parts, headpiece (residues, 1–73), rod (74–415) and tail piece (416–463). By treatment with proteases, such as chymotrypsin, head and tail may be easily hydrolysed, leaving a rod. The rod consists of three α helical regions flanked by nonhelical spacers: helix Ia (102–131), spacer 1 (132–145), helix Ib (146–242), spacer 2 (243–267) and

TABLE 2

AMINO ACID COMPOSITIONS OF CYTOSKELETAL PROTEINS OF VERTEBRATE SKELETAL MUSCLE (RESIDUES/1000 RESIDUES)

	Spectrin	*Desmin*	*Vimentin*	*β-Connectin*	*α-Actinin*	*Eu-actinin*	*Actin*	*ABP (Filamin)*	*Synemin*	*Amorphin*	*Z protein*	*Z-nin*	*34 kDa protein*	*Paratropomyosin*	*Tropomyosin*	*Nebulin*
Asp	103	97	128	96	114	89	91	87	97	117	102	104	111	56	64	111
Thr	38	56	62	76	53	73	12	62	46	43	51	69	47	17	17	70
Ser	63	71	89	60	53	62	56	68	55	46	57	100	88	25	28	62
Glu	200	176	185	111	185	102	107	114	178	115	116	128	186	155	150	128
Pro	14	36	27	74	37	42	49	71	28	46	58	45	0	0	0	30
Gly	89	64	27	74	46	75	86	117	67	72	63	87	85	10	9	52
Ala	92	92	63	65	84	83	82	74	96	86	49	74	103	76	76	76
Cys/2	—	3	2	2	—	8	13	5	3	—	5	—	—	—	—	8
Val	40	60	57	87	35	53	50	85	60	62	55	81	54	19	20	82
Met	19	18	20	12	31	40	43	13	15	20	19	—	—	12	12	7
Ile	32	41	31	60	62	77	77	44	53	44	64	54	49	23	24	61
Leu	103	94	107	66	92	79	75	62	106	100	117	75	95	66	68	79
Tyr	14	28	25	31	29	46	40	31	11	42	37	—	—	12	11	15
Phe	39	30	22	26	30	36	32	32	35	43	52	37	28	3	3	27
Lys	67	51	48	86	27	64	54	22	85	53	76	70	86	82	79	110
His	26	14	14	15	60	20	22	60	18	33	23	38	21	5	4	28
Arg	59	77	92	59	63	50	51	41	49	53	56	36	47	30	31	36
Ref.	*a*	36	*b*	11	111	88	88	*c*	*d*	111	114	118	118	124	124	59

[a] 240 K Fodrin from porcine brain.[10]
[b] Geisler, N. and Weber, K. (1981). Pig eye lens. *FEBS Lett.*, **125**, 253–6.
[c] Stossel, T. P. and Hartwig, J. H. (1976). Rabbit macrophage. *J. Cell Biol.*, **68**, 602–19.
[d] 220 K protein.[110]

helix 2 (268–408). It is of interest to note that the three α helical portions are highly conservative: the amino acid sequences are similar to each other in wool and epidermal keratins, vimentin, desmin, and glial intermediate filament protein.

The isolated 40 kDa rod piece of desmin is crosslinked by dimethylsuberimidate. The rod contains only one cysteine residue. Three species, monomer, dimer, and tetramer were detected by SDS-gel electrophoresis, suggesting that desmin protofilaments consist of dimeric organisation of double-stranded coiled coil rather than of the triple-stranded structures originally thought.

Izant and Lazarides first demonstrated the two isoforms of gizzard desmin (α, β) by two dimensional electrophoresis.[38] The isoelectric points of α- and β-desmins are approximately 5·65 and 5·70.[31] Porcine skeletal muscle desmin has a major isoform slightly more acidic than α-desmin.[33]

Polymerisation–depolymerisation

Cooke first observed that the desmin monomer, when dissolved in 6 M urea, easily polymerises into intermediate filaments on dialysis.[28] Synthetic filaments are formed by dialysing a desmin solution in 6 M urea against 150 mM NaCl, 10 mM imidazole buffer, pH 7·0.[32] The filaments are depolymerised into protofilaments, 2–2·5 nm wide and 30–70 nm long, by dialysis against 2 M urea for 1 h at pH 7·0 and 2 °C (or 8 mM Tris, pH 8·2 for 4 h).[39] Larger filaments, 3·5–5 nM in diameter, are also formed. It is very likely that the desmin protofilament consists of a tetramer of the 55 kDa subunit. The latter has a molecular weight of 200 kDa at low ionic strength.

It has remained obscure how the protofilaments are assembled into the 10 nm intermediate filaments. Steinert and his associates reported that the rate-limiting step for intermediate filament formation in keratin is the lateral association of two protofilaments.[40] Detailed investigation is needed to elucidate the polymerisation process of intermediate filaments. Unlike actin or tubulin, ATP or other nucleotides, are not required for intermediate filament formation.[29,31]

Role of Intermediate Filaments in Myofibrillar Structure

Granger and Lazarides isolated Z disc sheets from chicken leg muscle fibres and demonstrated that desmin is located at the periphery of Z disc.[41] Lazarides suggested that desmin filaments connect adjacent Z discs of neighbouring myofibrils whereby all the myofibrils are kept in register and thus form the familiar 'striation' of muscle fibres.[26,41]

The role of intermediate filaments in maintaining the continuity of a sarcomere has recently been emphasised.[42] It is suggested that several intermediate filaments run longitudinally from the periphery of a Z line to that of the adjacent one.[42] Tokuyasu has elegantly shown the presence of longitudinally-oriented desmin filaments over the surface of a cardiac myofibril using an immunoelectron microscopic technique.[43] The intermediate filaments have been detected in the intermyofibrillar spaces but not intramyofibrillar structures.[43] It should be pointed out that, since myofibrils may be stretched to twice their resting length by tension generation and will return to their initial state on release,[44] any intermediate filaments linking neighbouring Z lines must be elastic and must be slack in the resting state of myofibrils. Although the content in skeletal muscle is very small—only 0·3% of the total myofibrillar proteins—the desmin intermediate filaments must play an important role in the structural organisation of myofibrils.

3.2. Vimentin

Intermediate filaments of vimentin are characteristic of cells of mesenchymal origin and are also present in other types of cells.[45,46] Vimentin is named from the Latin *vimentum* (arrays of flexible rods).[46] Vimentin is an intermediate filament protein which is very similar to desmin. Its molecular weight is slightly larger, however (55–58 kDa), and its isoelectric point is somewhat more acidic (5·2–5·3).[27,31] The complete amino acid sequence of hamster lens vimentin has been predicted from the nucleotide sequence.[47] It consists of five regions: very basic NH_2 terminal domain (67 residues), α helix domains (135 and 145 residues) interspaced by a non-helical region (45 residues) and C terminal domain (55 residues). A neutral, calcium-activated proteinase specific to desmin and vimentin has been purified from Ehrlich ascites tumour cells[48] and its universal presence in all the vertebrate cells has been demonstrated.[49]

The coexistence of desmin and vimentin in Z lines of chicken skeletal muscle was demonstrated by Lazarides and his coworker.[41,50] It is also located at the periphery of Z disc in a collarlike arrangement. Vimentin filaments appear to function in a similar fashion to desmin filaments. The content of vimentin in chicken breast myofibrils is much less than that of desmin.[50] There has been no report on vimentin having been isolated from vertebrate skeletal muscle.

Interesting information was provided from the study of the synthesis and distribution of desmin and vimentin during differentiation of cultured muscle cells.[51] In chicken myoblasts, vimentin alone is synthesised; desmin synthesis is started at the onset of fusion of myoblasts. The rate of desmin synthesis is greatly increased over that of vimentin during myogenesis. In early myotubes both desmin and vimentin are present as free cytoplasmic filaments, and later the two proteins are incorporated into the Z band region of newly-assembled myofibrils.

4. ELASTIC PROTEIN. CONNECTIN (TITIN)

Huxley and Hanson first observed that continuity of Z lines is maintained even after removal of myosin and actin by successive extraction.[52] Sjöstrand claimed the presence of very thin filaments linking myosin filaments to Z lines and called the filament 'gap filament'.[53] On the other hand, Hoyle emphasised the presence of a 'T filament' running through neighbouring Z lines.[54]

In 1954, Natori demonstrated that there is an elastic structure in myofibrils, using skinned muscle fibres of the bullfrog.[44] Following Natori's procedure, the elastic component was shown to remain in ghost myofibrils after extraction of myosin, actin, and other proteins.[55] This insoluble elastic material consisted of a very high molecular weight protein (HMW) and a 43 kDa protein; the mixture was originally named connectin by the present author in 1977.[56] Connectin (HMW) was isolated by preparative SDS-gel electrophoresis[56] or gel filtration in the presence of 6 M guanidine-chloride.[57]

Two years later, Wang *et al.*[58] observed a very high molecular weight protein in a direct SDS-extract of chicken myofibrils and isolated it by gel filtration in the presence of SDS.[58] The protein was termed 'titin' (titin = very large). We have shown that titin is identical with the high molecular weight component of connectin.[59] Subsequently, the 43 kDa component of connectin was identified as denatured actin.[60] The term connectin has been used for the high molecular weight protein (also called titin). In 1981, we succeeded in preparing connectin in native state from chicken breast muscle.[61,62,63]

Preparation of Native Connectin

Based on our observations that connectin is partly solubilised, along with myosin, in 0·6 M KCl,[59] conditions in which connectin is extracted without myosin or actin were sought. Connectin was selectively extracted with 0·1 M phosphate buffer at pH 6·2–7·0 but not at pH 5·5–5·6. Thus, connectin was extracted with 0·1 M phosphate buffer, pH 6·6.[61] Chicken breast muscle myofibrils were well washed with 5 mM $NaHCO_3$, and extracted with 0·1 M phosphate buffer, pH 5·6, to remove denatured actin. Then, the residue was extracted several times at pH 6·6. As shown in Fig. 1, the filtrate consisted of almost pure connectin. The yield was as much as 200 mg starting from 50 g of muscle (4 % of the total myofibrillar proteins). This native connectin corresponded to β-connectin (titin 2). It seems that α-connectin (titin 1) is not solubilised, since it remains in the extracted residues.

Physicochemical Properties

As revealed by SDS-gel electrophoresis, there are two species of connectin, α and β (titins 1 and 2), in vertebrate striated muscle.[58,59] The molecular

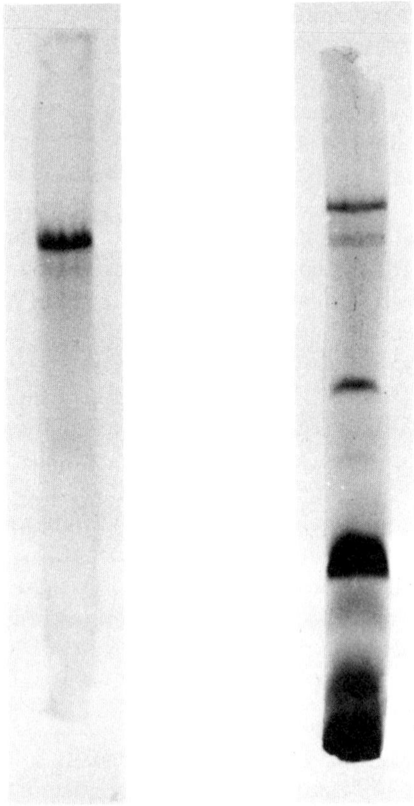

FIG. 1. An SDS-gel electrophoresis pattern of native connectin. 2% polyacrylamide gels. Note that native connectin corresponded to the lower band (β-connectin; titin 2) of the doublet in the extract of chicken breast muscle. Left, isolated native connectin; right, direct SDS extract of chicken breast muscle.

weights of α- and β-connectins are as high as 2800 kDa and 2100 kDa.*[64] Because of technical difficulty, these values must be regarded as tentative. It is very likely that β-connectin is a proteolytic product of α-connectin: in a direct SDS extract of muscle, α is predominant ($\alpha:\beta = 5:1$), in isolated myofibrils, the ratio of α is smaller ($\alpha:\beta = 1:1$) and isolated connectin is only β form. It appears that α-connectin is changed into β-connectin *post*

* Wang estimated the molecular weights of titins 1 and 2 to be 1800 kDa and 1200 kDa, respectively.[65] Crosslinked myosin heavy chains were used as MW markers in both investigations. Wang assumed the formation of dimers, trimers, tetramers, and so on. However, we have shown that dimers of heavy chains are crosslinked with each other.[64]

mortem.[66] There was no significant difference in the amino acid compositions between α-rich connectins (titins 1 and 2) directly obtained from muscle and native connectin (β-connectin).[61] Reactivity of α- and β-connectins against a set of monoclonal antibodies will settle this interesting problem. It is not thought that connectin consists of a single peptide having a chain weight of a few million, and possible subunits of connectin should be sought—as well as the nature of the chemical bonds between them. One of the candidates for the latter is hydroxylysinonorleucine, as in collagen.[67]

The molecular shape of β-connectin is a flexible rod, 300–400 nm long and 35–40 nm wide.[64] In a salt solution, β-connectin forms a loose network easily broken by a weak external force. When concentrated, connectin molecules are laterally associated to form elastic fibres. In salt solution, freshly prepared connectin has a sedimentation coefficient of 17 S ($S^{o}_{w,20}$) and there is a tendency for self-association (~30 S).

Interactions with Myosin and Actin

It is of physiological significance that connectin filaments can bind to those of both actin and myosin at a certain range of ionic strength ($<0{\cdot}16$). In the presence of 80 mM KCl and 5 mM phosphate buffer, pH 7·0, connectin made both myosin and actin filaments turbid, leading to flocculent precipitation.[68] There was no change in turbidity at KCl concentrations higher than 0·16 M KCl. Irregular bundles of actin filaments were observed under the electron microscope, and myosin filaments were entangled with each other to form huge aggregates. The interactions of connectin with actin and myosin filaments were reversible by changing KCl concentrations. Therefore, the nature of the interactions appears to be electrostatic. It should be mentioned that, in the case of connectin and actin filaments, a weak mechanical force greatly hindered the interaction.

Connectin appreciably enhances the onset of Mg-ATP-induced superprecipitation of actomyosin.[69] At the same time, ATPase activity was accelerated, but to a lesser extent than α-actinin. The calcium-dependent ATPase activity of the sensitised actomyosin, in the presence of the troponin–tropomyosin system, was greatly activated by connectin in the presence of 0·2 mM $CaCl_2$. On the other hand, the Ca-activated ATPase activity in the presence of 5 mM $CaCl_2$ was not affected at all. The enhancement of superprecipitation of actomyosin by connectin may be due to artefacts as by α-actinin due to bundle formation of actin filaments.

Location of Connectin in Myofibrils

Indirect immunofluorescence studies have revealed that connectin is mainly concentrated in the A–I junction area, abundantly present in both sides of myosin filaments (the middle portion is connectin-poor), and is present in small amounts in the A band.[58,59,70] Connectin seems to be not present in Z lines. However, when myosin and actin are solubilised in 0·6 M KI, connectin is accumulated in either sides of Z lines together with actin and myosin.[56,58] The latter is evidently an artefact.[41]

There are entanglements of very thin filaments, 2 nm in diameter, in thin sections of muscle residues.[71] In quick-freeze, deeply-etched, replica specimens, these thin filaments were very similar to connectin filaments in artificially made fibres of native connectin.[72] Very thin filaments have been clearly shown in the A–I junction area of extremely stretched frog muscle fibres.[73] Indirect immunofluorescence microscopy strongly suggests that these filaments are connectin filaments. The filaments linked myosin filaments eventually to the Z lines. Therefore, connectin filaments are actually Sjöstrand's gap filament,[53] as revived by Locker and his coworkers.[74]

Physiological Role of Connectin

Initially, it was claimed that connectin is simply responsible for longitudinal continuity of a sarcomere linking neighbouring Z lines.[55,56] However, as work has progressed, the situation has become more complicated and opened a new field in structural integrity of striated muscle myofibrils, although the original idea still holds true.

First, it is definitely true that elastic connectin filaments link myosin filaments to Z lines, forming gap filaments, as Sjöstrand[53] and Locker[74] proposed. Morphological evidence shows that there are connectin filaments attaching to both sides of myosin filaments,[73] and furthermore, in highly stretched myofibrils, myosin filaments in an A band are pulled away randomly toward either side of adjacent Z lines.[73] Generation of passive tension upon stretch is linearly decreased as myosin is extracted, and on nearly complete removal of myosin, only one tenth of the passive tension is developed.[75] This may be explained by successive dissolution of connectin bound to myosin filaments. In addition, connectin nets at the A–I junction area may serve to keep myosin filaments at the centre of a sarcomere.

Here a question arises. Do connectin filaments fuse into a smaller number of filaments directly running through Z lines? From the immunofluorescence study, connectin is only faintly present in I bands.[59,73] It is also possible that connectin filaments bind to actin filaments. Tips of actin filaments are free judging from movement of actin filaments in extremely stretched fibres by an application of electric field.[76] Another possibility is the binding of connectin filaments to other elastic filaments linking to Z lines. One of the candidates is nebulin as described by Wang.[77] It should be pointed out, however, that nebulin does not directly bind to Z lines,[77,78] but to actin filaments or to other Z line-connecting proteins. All these interesting problems wait for further experiments in the near future.

Finally, one should discuss the longitudinal linking of adjacent Z lines. There are few connectin-like thin filaments running through Z lines in highly stretched myofibrils.[73] As mentioned above, most of connectin filaments are concentrated on two regions in both sides of myosin filaments. It is certain that a few desmin intermediate filaments run through Z lines on the periphery of a sarcomere.[43] However, judging from the passive elasticity, the intermediate filaments must be very slack in resting state. Figure 2 illustrates our present understanding of cytoskeletal elastic structure of a sarcomere. The simple model suggests the sole role of connectin in passive elasticity of the myofibril, whereas, in the complicated model, two more elastic proteins are involved. Future work will elucidate the real mechanism.

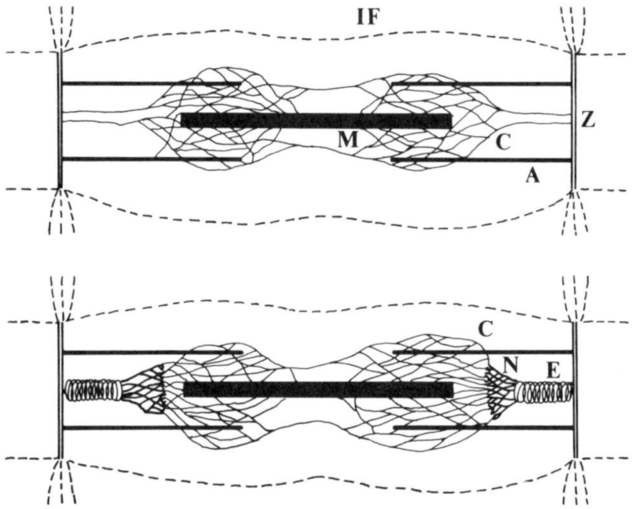

FIG. 2. Diagrams of cytoskeletal framework of a sarcomere of vertebrate striated muscle. Upper—simple model; connectin nets (C) are responsible for elasticity. Lower—complicated model; it is assumed that connectin nets (C) are linked to other elastic nets (N, possible candidate, Wang's nebulin,[75,76] and the latter nets are further connected to elastic filaments (E, unknown) attached to Z lines. A, actin filaments; M, myosin filaments; C, connectin filaments; N, nebulin(?); E, unknown elastic filaments; IF, desmin intermediate filaments; Z, Z lines.

5. Z DISC PROTEINS

Z disc (Z line) is an important structure, placing actin filaments into register in myofibrils as well as uniformly segmenting the myofibrils into the structural units, sarcomeres. There have been a number of ultrastructural studies of Z discs of vertebrate skeletal muscle which demonstrate their highly ordered tetragonal lattice array.[79] There are two categories of the Z disc proteins: one group contains desmin, vimentin, synemin, and actin-binding protein (ABP; filamin); these encircle the periphery of the Z lines. The other group (actin, α-actinin, and Z protein), is located in the interior of Z discs.[41] Unfortunately, little has been known of the structural organisation of those proteins. The intermediate filament proteins desmin and vimentin were considered above.

5.1. α-Actinin

α-Actinin was discovered by the Ebashis and the present writer as a contraction-promoting factor in 1964.[80] α-Actinin gelates an F-actin solution by forming irregular bundles of actin filaments.[81,82] α-Actinin (95 kDa) is a rod (4 × 40 nm) consisting of a dimer.[83] The exchange of ADP bound to F-actin with free ATP is enhanced by α-actinin, suggesting that α-actinin modulates the dynamic state of actin filaments.[84] It is located in Z lines, contributing as much as 50% of the total proteins of Z line.[83] Granger and Lazarides[41] showed that α-actinin is uniformly distributed in the interior of isolated Z discs. Therefore, it is thought that α-actinin is a cementing substance to fix actin filaments at the Z line.[5] Z lines become faint either by extraction with a low ionic strength solution or by a protease treatment (for example, Ca-activated neutral protease.[85] It is not clear whether a protein is involved in the deposit of α-actinin in Z lines or not.

Recently, Takahashi and his coworkers have observed an interesting phenomenon:[86] treatment of rabbit psoas myofibrils with 0·1 mM Ca^{2+}, in the presence of an inhibitor (E 64 or calpastatin) of calcium-activated protease, resulted in weakening of Z bands; and when homogenised, fragmentation of myofibrils occurred. However, α-actinin was not released in this case unlike the action of proteases. This weakening of Z disc is regarded as resulting from separation of double lattice structures of Z disc. Takahashi believes that the Z disc lattice structure consists of α-actinin itself.[87] The latter view has been also supported by other workers (see below, Amorphin). It is desirable to get direct evidence, for example, by immunoelectron microscopic observations.

In smooth muscle, α-actinin is located in cytoplasmic and plasma membrane dense bodies.[88] In non-muscle cells α-actinin is located in adhesion plaques of the plasma membrane, actin-filament-membrane attachment area, and also in stress fibres in a patchy pattern.[89] It is of interest to note that α-actinin induces polymerisation of G-actin bound to profilin, a cytoplasmic inhibitor of actin polymerisation.[90]

5.2. Eu-actinin

It was long thought that polymerisable G-actin cannot be extracted from myosin-removed myofibrils without acetone treatment. However, it turned out that a water extract of myofibrils contains ATPases that denature G-actin by splitting the bound ATP and therefore, active G-actin can be extracted from myofibrils by an ATP solution.[91] During the course of this study, Kuroda observed that a new actin-associated protein is extracted at pH 6·5 while actin is solubilised at neutral pH.[92] Eu-actinin, thus designated, has not only the same molecular weight (43 000) as actin but also an amino acid composition very similar to actin (Table 2). It is located in Z lines. It is not as yet determined whether eu-actinin is present in the interior of the Z line or not. Eu-actinin interacts with actin and bundles of actin filaments are formed by eu-actinin.[93] Unlike α-actinin, however, eu-actinin inhibits the onset of superprecipitation of actomyosin. Interestingly, α-actinin releases the inhibiting action of eu-actinin. The content of eu-actinin is approximately 10 mg per 100 g of chicken breast muscle.

5.3. Actin Binding Protein (ABP, Filamin)

Actin-binding protein (ABP) (MW = 270 kDa) was originally discovered by Hartwig and Stossel in macrophage.[94] A similar actin-gelating protein was isolated from chicken gizzard and termed filamin.[95] Although filamin has been used widely in the literature, the term ABP (e.g. gizzard ABP) is adopted here (according to Stossel).[96] At physiological ionic strength, ABP exists as a dimer (MW = 540 kDa) 16·2 nm long. One ABP dimer can bind up to 14 actin monomers into an actin filament, binding occurring at 40 nm intervals. ABP dimers crossbridge two actin filaments resulting in gelation of F-actin in solution.[97] The gelating action of ABP is hindered by tropomyosin.[98]

In chicken gizzard, there is a calmodulin-binding protein caldesmon (MW = 150 kDa) that interacts with F-actin.[99,100] Caldesmon interferes with ABP-induced gelation of F-actin, irrespective of Ca^{2+} concentrations. It is to be noted that the amount of the ABP bound to actin filaments is not changed by caldesmon.[100] In the presence of calmodulin, the gelation is only inhibited in the absence of Ca^{2+}, since, in the presence of Ca^{2+}, a calmodulin–caldesmon complex is formed leaving from the actin filaments. Thus, cytoskeletal structures gain Ca^{2+} sensitivity.[101]

The presence of ABP in the periphery of Z lines of chicken breast muscle was clearly demonstrated by an indirect immunofluorescence technique.[102] Bechtel isolated ABP from chicken skeletal muscle.[103]

An interesting feature of the appearance of Z disc proteins has been found during chicken skeletal myogenesis *in vitro*.[102,104] ABP is associated with α-actinin-containing actin filaments in myoblasts and, one day after fusion, ABP disappears. Several days after the appearance of α-actinin-associated Z lines ABP reappears in the Z lines, together with vimentin- and desmin-intermediate filaments. Analysis by peptide mapping and isoelectric point have suggested that the ABP molecules present in myoblasts and myotubes are different species.[104]

5.4. Synemin

When desmin is prepared from chicken gizzard, it is found that a high molecular weight protein (MW = 230 kDa) is closely associated with the desmin filament.[105] The content was only 1 % of desmin. By preparative SDS-gel electrophoresis, the protein was separated from desmin, and antisera raised in a rabbit. The protein was named synemin ('with-filament', in Greek). Synemin was shown to coexist with desmin and vimentin filaments at the periphery of Z lines in chicken skeletal muscle. The three kinds of proteins coexist in the cytoplasm of cultured myotubes, and within a week after myoblast fusion, they attach to the Z lines. Lazarides and his associate regard synemin as one of the intermediate filament proteins,[104] because, in avian erythrocytes, synemin is present as an intermediate filament together with vimentin.[106,107] It should be pointed out that the MW of synemin is eight-fold that of other intermediate filament proteins.

On the other hand, Fukazawa and his coworkers claimed that a high molecular weight protein (MW = 220 kDa) is a constituent of Z lines and could be solubilised by 1 mM $CaCl_2$[108] or by treatment with calcium-activated neutral protease.[109] The 220 kDa protein was isolated from rabbit and chicken skeletal muscle by gel filtration in the presence of 6 M urea or 0·1 % SDS, and its location in Z lines was demonstrated.[110] The amino acid composition of this protein was different from those of vimentin, desmin, α-actinin, and ABP (Table 2). Judging from the similar values for MW, the 220 kDa protein is presumably the same as synemin.

5.5. Amorphin

In 1982, Chowrashi and Pepe[111] reported that a 85 kDa protein, called amorphin, is located in Z lines of chicken breast muscle. Amorphin can be extracted by buffered sucrose solution leaving α-actinin behind, which the authors thought to constitute the lattice structure filaments. Amorphin contributes to the density of Z line.

5.6. Z Protein

The lattice-forming proteins of the Z line in vertebrate skeletal muscle remained obscure for a long time. Ohashi and Maruyama were able to isolate Z protein (MW = 55 kDa) which presumably constitutes the lattice structure of the Z lines.[112,113,114] Myosin-free myofibrils of chicken breast muscle are extracted with 0·6 M KI for 24 h together with actin. Z protein is purified from the extract by ammonium sulphate fractionation (20–40 %) followed by DEAE-Sephacel chromatography. The MW of Z protein is the same as that of desmin, but the two proteins are distinguishable by immunoreactivity and amino acid composition. When this protein was kept standing at 5 °C for a few days in 1 mM $NaHCO_3$, a fine precipitation occurred. Lattice structures, with a periodicity of 8 nm, were observed under the electron microscope. The Z protein was shown to be located in the interior of the Z disc.[114] So far, Z protein does not appear to interact with α-actinin or actin. There must be some cementing protein to locate actin filaments in the lattice structure.

5.7. Z-nin

Suzuki and his associates observed that some proteins released from Z lines with calcium-activated neutral protease[115] can reconstitute the Z lines of myofibrils that had been damaged by prolonged extraction with 2 mM Tris-HCl, pH 7·6.[116] The likely identity of the proteins required for the reconstitution of Z lines were those with the chain weights of 300–400 kDa, 100 kDa (α-actinin) and 34 kDa (tropomyosin-like protein).[117] The high molecular weight protein (300–400 kDa) was named Z-nin.[117] The localisation of Z-nin in Z lines has been demonstrated by indirect immunofluorescence[118] and immunoelectron microscopic[119] observations. However, the fine structure of this interesting protein in the Z lines has remained obscure. Z-nin differs from synemin in the amino acid composition (Table 2). Since the native form of Z-nin has been purified,[117] it is desirable to investigate its molecular shape and possible interactions with other proteins such as actin and α-actinin. Also it is necessary to determine the precise chain weight of Z-nin. It should be easy to estimate a chain weight of 300–400 kDa using 2·5 % SDS polyacrylamide gels.

Suzuki and his coworkers reported that a 34 kDa protein is also necessary to reconstitute the Z lines.[119,120] The nature of this tropomyosin-like protein has not yet been elucidated. The amino acid composition of the 34 kDa protein is similar to that of tropomyosin lacking proline (Table 2). Tropomyosin itself has never been shown to be located in Z lines.

6. OTHER CYTOSKELETAL PROTEINS

6.1. Nebulin

When a piece of chicken breast muscle is directly dissolved in an SDS solution and subjected to gel electrophoresis, an unknown high molecular weight protein, called band 3, was detected in addition to connectin (titin) bands.[58] The molecular weight was estimated to be 500 kDa. In our hands, it was approximately 800 kDa (Kimura, S., Yoshidomi, H., and Maruyama, K., unpublished). This protein was isolated by gel filtration in the presence of SDS, and the antibodies were raised in a rabbit.[77] It is named nebulin, because of its location at the N_2 line, the nebulous striation in either side of the Z line.[121] When a sarcomere is stretched, the N_2 line is moved in such a way that the proportional distance to the adjacent M line and Z line is kept constant suggesting that the N_2 line is attached to the elastic filaments extending from the M line to the Z line. The latter could be connectin filaments (cf. Fig. 2). It would be highly desirable to obtain nebulin in a native state so that its interactions with actin and connectin be investigated.

6.2 Paratropomyosin

Takahashi and his coworkers observed that the amount of myosin and actin filaments released by the addition of ATP was only 5% of the total proteins in the myofibrils prepared from freshly excised muscle, whereas it was as much as 50% in those from rigor muscle.[122] Even with fresh muscle, when myofibrils were treated with 0·1 mM $CaCl_2$, the extent of dissociation of myosin and actin filaments increased as with rigor muscle.[123] A protein called paratropomyosin (MW = 34 000) was isolated from the 0·1 mM Ca extract.[124] Paratropomyosin inhibits the actin-myosin interactions in the presence of ATP. It is assumed that paratropomyosin originally located at the tips of myosin filaments are translocated to actin filaments during post-mortem changes to weaken myosin and actin interactions.

7. CONCLUDING REMARKS

Cytoskeletal aspects of vertebrate striated muscle have been somewhat unpopular with researchers, who are mainly concerned with the attractive contractile and regulatory mechanisms of muscle contraction. Thus, the fine structure of even the Z lines has not yet been elucidated satisfactorily, although a number of possible constituting proteins have been discovered during the past 15 years. Ironically, stimulated by a rapid progress in the work of cytoskeletons of nonmuscle cells, cytoskeletal structure and function of striated muscle have recently attracted attention of an increasing number of researchers. The elastic network of connectin (titin) is one of such topics. There remain many problems to be solved, as shown in Fig. 2.

The cytoskeletal framework of striated muscle is of very great significance from the viewpoint of meat science. One of the factors determining meat quality is its texture. It greatly depends on the post-mortem changes in the cytoskeletal structures of myofibrils including myosin and actin filaments. It is generally accepted that activated proteases digest myofibrillar structural proteins *post mortem*.[115] There are, however, other events: elevation in free Ca^{2+} concentration to 0·1–0·2 mM in post-mortem muscle fibres. Takahashi[90] emphasises the Ca^{2+}-induced weakening (tenderisation) of myofibrillar structures during post-mortem storage: the weakening of Z discs, the weakening of rigor linkages formed between actin and myosin by dislocated paratropomyosin, and the breakdown of elastic connectin networks.

The myofibrillar cytoskeletal system constitutes an interesting and fruitful field of future research, not only from the basic point of view but also from the viewpoint of applied science, especially meat science.

8. ACKNOWLEDGEMENT

The writer is greatly indebted to his colleague, Dr S. Kimura for her assistance in the preparation of the present review.

9. REFERENCES

1. Ebashi, S., Maruyama, K. and Endo, M. (Eds) (1980). *Muscle contraction. Its regulatory mechanisms*, Japan Sci. Soc. Press, Tokyo/Springer Verlag, Berlin.
2. Korn, E. D. (1978). *Proc. Natl. Acad. Sci. USA*, **75**, 588.
3. Ebashi, S. (1980). *Proc. Roy. Soc. London*, **207B**, 259.
4. Poste, G. and Nicolson, G. L. (Eds) (1981). *Cytoskeletal elements and plasma membrane organization. Cell surface reviews*, Vol. 7, North-Holland, Amsterdam.

5. Obinata, T., Maruyama, K., Sugita, H., Kohama, K. and Ebashi, S. (1981). *Muscle & Nerve*, **4**, 456.
6. Yates, L. D. and Greaser, M. L. (1983). *J. Mol. Biol.*, **168**, 123.
7. Branton, D., Cohen, C. M. and Tyler, J. (1981). *Cell*, **24**, 24.
8. Goodman, S. R., Zagon, I. S. and Kulikowski, R. R. (1981). *Proc. Natl. Acad. Sci. USA*, **78**, 7570.
9. Tsukita, S., Tsukita, S., Ishikawa, H., Sato, S. and Nakao, M. (1981). *J. Cell Biol.*, **90**, 70.
10. Glenney, J. R. Jr., Glenney, P. and Weber, K. (1982). *J. Biol. Chem.*, **257**. 9781.
11. Geiger, B. (1979). *Cell*, **18**, 193.
12. Burridge, K. and Feramisco, J. R. (1980). *Cell*, **19**, 587.
13. Geiger, B., Dutton, A. H., Tokuyasu, K. T. and Singer, S. J. (1981). *J. Cell Biol.*, **91**, 614.
14. Singer, I. I. and Paradiso, P. (1981). *Cell*, **24**, 481.
15. Feramisco, J. R. and Burridge, K. (1980). *J. Biol. Chem.*, **255**, 1194.
16. Endo, T. and Masaki, T. (1982). *Biochem. Biophys. Res. Comm.*, **107**, 1467.
17. Isenberg, G., Leonard, K. and Jockusch, B. (1982). *J. Mol. Biol.*, **158**, 231.
18. Geiger, B. (1982). *J. Mol. Biol.*, **159**, 685.
19. Jockusch, B. M. and Isenberg, G. (1981). *Proc. Natl. Acad. Sci. USA*, **78**, 3005.
20. Wilkins, J. A. and Lin, S. (1982). *Cell*, **28**, 83.
21. Pardo, J. V., Siliciano, J. and Craig, S. W. (1983). *Proc. Natl. Acad. Sci. USA*, **80**, 1008.
22. Pardo, J. V., Siliciano, J. and Craig, S. W. (1983). *J. Cell Biol.*, **97**, 1081.
23. Otto, J. J. (1983). *J. Cell Biol.*, **97**, 1283.
24. Lazarides, E. and Weber, K. (1974). *Proc. Natl. Acad. Sci. USA*, **73**, 4344.
25. Ishikawa, H., Bischoff, R. and Holtzer, H. (1969). *J. Cell Biol.*, **43**, 312.
26. Lazarides, E. (1980). *Nature*, **283**, 249.
27. Zackroff, R. V., Steinert, P. M., Aynardi Whitman, M. and Goldman, R. D. (1981). In: *Cell surface reviews* (Poste, G. and Nicolson, G. L. (Eds)), Vol. 7, North-Holland, Amsterdam, p. 55.
28. Cooke, P. (1976). *J. Cell Biol.*, **68**, 539.
29. Small, J. V. and Sobieszek, A. (1977). *J. Cell Sci.*, **23**, 243.
30. Lazarides, E. and Hubbard, B. D. (1976). *Proc. Natl. Acad. Sci. USA*, **73**, 4344.
31. Lazarides, E. and Granger, B. L. (1982). In: *Methods in enzymology* (Frederiksen, D. W. and Cunningham, L. W. (Eds)), Vol. 85B, Academic Press, New York, p. 488.
32. Huiatt, T. W., Robson, R. M., Arakawa, N. and Stromer, M. H. (1980). *J. Biol. Chem.*, **255**, 6981.
33. O'Shea, J. M., Robson, R. M., Hartzer, M. K., Huiatt, T. W., Rathbun, W. E. and Stromer, M. H. (1981). *Biochem. J.*, **195**, 345.
34. Stigbrand, T., Eriksson, A. and Thornell, L. E. (1979). *Biochim. Biophys. Acta*, **577**, 52.
35. Steinert, P. M., Idler, W. W. and Goldman, R. D. (1980). *Proc. Natl. Acad. Sci. USA*, **77**, 4534.
36. Geisler, N., Kaufmann, E. and Weber, K. (1982). *Cell*, **30**, 277.

37. GEISLER, N. and WEBER, K. (1982). *EMBO J.*, **1**, 1649.
38. IZANT, J. G. and LAZARIDES, E. (1977). *Proc. Natl. Acad. Sci. USA*, **74**, 1450.
39. STROMER, M. H., HUIATT, T. W., RICHARDSON, F. L. and ROBSON, R. M. (1981). *Eur. J. Cell Biol.*, **25**, 136.
40. STEINERT, P. M., IDLER, W. W. and ZIMMERMAN, S. B. (1976). *J. Mol. Biol.*, **108**, 547.
41. GRANGER, B. L. and LAZARIDES, E. (1978). *Cell*, **15**, 1253.
42. WANG, K. and RAMIREZ-MITCHELL, R. (1983). *J. Cell Biol.*, **96**, 562.
43. TOKUYASU, K. T. (1983). *J. Cell Biol.*, **97**, 562.
44. NATORI, R. (1954). *Jikeikai Med. J.*, **1**, 119.
45. HYNES, R. O. and DESTREE, A. I. (1978). *Cell*, **13**, 151.
46. FRANKE, W. W., SCHMID, E., OSBORN, M. and WEBER, K. (1978). *Proc. Natl. Acad. Sci. USA*, **75**, 5034.
47. QUAX-JEUKEN, Y. E. F. M., QUAX, W. J. and BLOEMENDAL, H. (1983). *Proc. Natl. Acad. Sci. USA*, **80**, 3548.
48. NELSON, W. J. and TRAUB, P. (1982). *J. Biol. Chem.*, **257**, 5544.
49. NELSON, W. J. and TRAUB, P. (1982). *J. Cell Sci.*, **57**, 25.
50. GRANGER, B. L. and LAZARIDES, E. (1979). *Cell*, **18**, 1053.
51. GARD, D. L. and LAZARIDES, E. (1980). *Cell*, **19**, 263.
52. HUXLEY, H. E. and HANSON, J. (1954). *Nature*, **173**, 971.
53. SJÖSTRAND, F. S. (1962). *J. Ultrastruct. Res.*, **7**, 225.
54. HOYLE, G., MCNEIL, A., WALCOTT, B. and SELVERSTON, A. (1968). *Symp. Biol. Hung.*, **8**, 34.
55. MARUYAMA, K., NATORI, R. and NONOMURA, Y. (1976). *Nature*, **262**, 58.
56. MARUYAMA, K., MATSUBARA, S., NATORI, R., NONOMURA, Y., KIMURA, S., OHASHI, K., MURAKAMI, F., HANDA, S. and EGUCHI, G. (1977). *J. Biochem.*, **82**, 317.
57. KING, N. L. and KURTH, L. (1980). In: *Fibrous proteins, scientific, industrial and medical aspects* (Parry, D. A. D. and Creamer, L. K. (Eds)), Vol. 2, Academic Press, London, p. 57.
58. WANG, K., MCCLURE, J. and TU, A. (1979). *Proc. Natl. Acad. Sci. USA*, **76**, 3698.
59. MARUYAMA, K., KIMURA, S., OHASHI, K. and KUWANO, Y. (1981). *J. Biochem.*, **89**, 701.
60. MARUYAMA, K., YAMADA, N., IKEYA, H. and KIMURA, S. (1983). In: *Muscular dystrophy: Biomedical aspects* (Ebashi, S. and Ozawa, E. (Eds)), Japan Sci. Soc. Press, Tokyo/Springer Verlag, Berlin, p. 201.
61. KIMURA, S. and MARUYAMA, K. (1983). *J. Biochem.*, **94**, 2083.
62. MARUYAMA, K. and KIMURA, S. (1981). *Current Res. Muscular Dystrophy, Japan*, **3**, 47.
63. KIMURA, S., SAWADA, H. and MARUYAMA, K. (1982). *Biophysics* (*Japan*), **22**, S34.
64. MARUYAMA, K., KIMURA, S., YOSHIDOMI, H., SAWADA, H. and KIKUCHI, M. (1984). *J. Biochem.*, **95**, 1423.
65. WANG, K. (1982). In: *Methods in enzymology* (Frederiksen, D. W. and Cunningham, L. W. (Eds)), Vol. 85, Academic Press, New York, p. 264.
66. SEKI, N. and WATANABE, T. (1984). *J. Biochem.*, **95**, 1161.

67. Fujii, K. and Maruyama, K. (1981). *Biochem. Biophys. Res. Comm.*, **104**, 633.
68. Kimura, S. and Maruyama, K. (1983). *Biomed. Res.*, **4**, 607.
69. Kimura, S., Maruyama, K. and Huang, Y. P. (1984). *J. Biochem.*, **96**, 499.
70. Maruyama, K., Kimura, S., Toyota, N. and Ohashi, K. (1980). In: *Fibrous proteins, scientific, industrial and medical aspects* (Parry, D. A. D. and Creamer, L. K. (Eds)), Vol. 2, Academic Press, London, p. 33.
71. Toyoda, N. and Maruyama, K. (1978). *J. Biochem.*, **84**, 239.
72. Sawada, H., Maruyama, K. and Kimura, S. (1983). *Biomed. Res.*, **4**, 603.
73. Maruyama, K., Kimura, S., Ohashi, K., Higuchi, H. and Umazume, Y. (1984). *J. Cell Biol.*, **99**, 1391.
74. Locker, R. H. and Daines, G. J. (1980). In: *Fibrous proteins, scientific, industrial and medical aspects* (Parry, D. A. D. and Creamer, L. K. (Eds)), Vol. 2, Academic Press, London, p. 43.
75. Higuchi, H. and Umazume, Y. (1983). *Biophysics* (*Japan*), **23**, S79.
76. Umazume, Y. and Fujime, S. (1975). *Biophys. J.*, **15**, 163.
77. Wang, K. and Williamson, C. L. (1980). *Proc. Natl. Acad. Sci. USA*, **77**, 3254.
78. Wang, K. (1982). In: *Muscle development* (Pearson, M. L. and Epstein, F. (Eds)), Cold Spring Harbor Laboratory, New York, p. 339.
79. Goldstein, M. A., Schroeter, J. P. and Sass, R. L. (1982). *J. Muscle Res. Cell Motility*, **3**, 333.
80. Ebashi, S., Ebashi, F. and Maruyama, K. (1964). *Nature*, **203**, 645.
81. Maruyama, K. and Ebashi, S. (1965). *J. Biochem.*, **58**, 13.
82. Pudlubnaya, Z. A., Tskhovrebova, L. A., Zaalishvili, M. M. and Stefanenko, G. A. (1975). *J. Mol. Biol.*, **92**, 357.
83. Suzuki, A., Goll, D. E., Singh, I., Allen, R. E., Robson, R. M. and Stromer, M. H. (1976). *J. Biol. Chem.*, **251**, 6860.
84. Craig-Schmidt, M. C., Robson, R. M., Goll, D. E. and Stromer, M. H. (1981). *Biochim. Biophys. Acta*, **670**, 9.
85. Reddy, M. K., Etlinger, J. D., Rabinowitz, M., Fischman, D. A. and Zak, R. (1975). *J. Biol. Chem.*, **250**, 4278.
86. Hattori, A. and Takahashi, K. (1982). *J. Biochem.*, **92**, 381.
87. Takahashi, K., Hattori, A., Nakamura, F. and Yamanoue, M. (1983). *Proc. Vth World Conf. Animal Production*, Tokyo.
88. Fay, F. S., Fujiwara, K., Rees, D. D. and Fogarty, K. E. (1983). *J. Cell Biol.*, **96**, 783.
89. Jockusch, B. (1979). In: *Cell motility: molecules and organization* (Hatano, S., Ishikawa, H. and Sato, H. (Eds)), Univ. Tokyo Press, Tokyo, p. 121.
90. Blikstad, I., Eriksson, S. and Carlsson, L. (1980). *Eur. J. Biochem.*, **109**, 317.
91. Kuroda, M. (1982). *J. Biochem.*, **92**, 1863.
92. Kuroda, M., Tanaka, T. and Masaki, T. (1981). *J. Biochem.*, **89**, 297.
93. Kuroda, M. and Masaki, T. (1985). *J. Biochem.*, in press.
94. Hartwig, J. H. and Stossel, T. P. (1975). *J. Biol. Chem.*, **250**, 5699.
95. Wang, K. and Singer, S. J. (1977). *Proc. Natl. Acad. Sci. USA*, **74**, 2021.
96. Stossel, T. P., Hartwig, J. H., Yin, H. L. and Davies, W. A. (1979). In: *Cell motility: Molecules and organization* (Hatano, S., Ishikawa, H. and Sato, H. (Eds)), Univ. Tokyo Press, Tokyo, p. 189.

97. HARTWIG, J. H. and STOSSEL, T. P. (1981). *J. Mol. Biol.*, **145**, 563.
98. MARUYAMA, K. and OHASHI, K. (1978). *J. Biochem.*, **84**, 1017.
99. SOBUE, K., MURAMOTO, Y., FUJITA, M. and KAKIUCHI, S. (1981). *Proc. Natl. Acad. Sci. USA*, **78**, 5652.
100. MARUYAMA, K., SOBUE, K., MORIMOTO, K., KANDA, K. and KAKIUCHI, S. (1985). *J. Biochem.*, **97**, 1517.
101. KAKIUCHI, S. and SOBUE, K. (1983). *Trends in Biochem. Sci.*, **8**, 59.
102. GOMER, R. H. and LAZARIDES, E. (1981). *Cell*, **23**, 524.
103. BECHTEL, P. J. (1979). *J. Biol. Chem.*, **254**, 1755.
104. GOMER, R. H. and LAZARIDES, E. (1983). *J. Cell Biol.*, **96**, 321.
105. GRANGER, B. L. and LAZARIDES, E. (1980). *Cell*, **22**, 727.
106. GRANGER, B. L. and LAZARIDES, E. (1982). *Cell*, **30**, 263.
107. GRANGER, B. L., REPASKY, E. A. and LAZARIDES, E. (1982). *J. Cell Biol.*, **92**, 299.
108. MUGURUMA, M., YAMADA, M. and FUKAZAWA, T. (1978). *Biochim. Biophys. Acta*, **532**, 71.
109. MUGURUMA, M., MUGURUMA, Y. and FUKAZAWA, T. (1980). *J. Biochem.*, **88**, 145.
110. MUGURUMA, M., KOBAYASHI, K., FUKAZAWA, T., OHASHI, K. and MARUYAMA, K. (1981). *J. Biochem.*, **89**, 1981.
111. CHOWRASHI, P. K. and PEPE, F. A. (1982). *J. Cell Biol.*, **94**, 565.
112. OHASHI, K. and MARUYAMA, K. (1979). *J. Biochem.*, **85**, 1103.
113. OHASHI, K. and MARUYAMA, K. (1980). In: *Muscle contraction: Its regulatory mechanisms* (Ebashi, S., Maruyama, K. and Endo, M. (Eds)), Japan Sci. Soc. Press, Tokyo, p. 497.
114. OHASHI, K., MIKAWA, T. and MARUYAMA, K. (1982). *J. Cell Biol.*, **95**, 85.
115. GOLL, D. E., OTSUKA, Y., NAGANIS, P. A., SHANNON, J. D., SATHE, S. K. and MUGURUMA, M. (1983). *J. Food Biochem.*, **7**, 137.
116. SUZUKI, A., SAITO, M., IWAI, H. and NONAMI, Y. (1978). *Agr. Biol. Chem.*, **42**, 2117.
117. SUZUKI, A., SAITO, M., OKITANI, A. and NONAMI, Y. (1981). *Agr. Biol. Chem.*, **45**, 2535.
118. SUZUKI, A. and NONAMI, Y. (1982). *Agr. Biol. Chem.*, **46**, 1103.
119. SUZUKI, A., SAITO, M. and NONAMI, Y. (1983). *Agr. Biol. Chem.*, **47**, 171.
120. SUZUKI, A. and NONAMI, Y. (1982). *Agr. Biol. Chem.*, **46**, 1977.
121. FRANZINI-ARMSTRONG, C. (1970). *Tissue and Cell*, **2**, 327.
122. TAKAHASHI, K., NAKAMURA, F. and INOUE, A. (1981). *J. Biochem.*, **89**, 321.
123. TAKAHASHI, K., NAKAMURA, F. and OKAMOTO, M. (1982). *J. Biochem.*, **92**, 809.
124. TAKAHASHI, K., NAKAMURA, F., HATTORI, A. and YAMANOUE, M. (1985). *J. Biochem.*, **97**, 1043.

ADDENDUM

Recent re-investigations of vinculin have shown that its capping action was due to a minor contaminant in routine vinculin preparations.[125]

Preparation and molecular characteristics of native connectin (titin) have been recently reported.[126,127] The results are almost the same as described in this chapter. On the other hand, studies using monoclonal antibodies against connectin (titin) have suggested that the connectin filaments consist of single giant molecules extending from each half of the A band toward adjacent Z lines.[128] Wang has presented a view that connectin (titin) is linked to nebulin meshwork,[129] while we have shown immunoelectron microscopic evidence that it is indeed linked to Z lines.[130] Recently, a comprehensive review on gap filament (connectin) including its role in meat tenderness has been published by R. H. Locker.[131]

125. EVANS, R. R., ROBSON, R. M. and STROMER, M. H. (1984). *J. Biol. Chem.*, **259**, 3916.
126. WANG, K., RAMIREZ-MITCHELL, R. and PALTER, D. (1984). *Proc. Natl. Acad. Sci. USA*, **81**, 3685.
127. TRINICK, J., KNIGHT, P. and WHITING, A. (1984). *J. Mol. Biol.*, **180**, 331.
128. WANG, K. (1985). Abst. Int. Congress of Cell Biology, Tokyo.
129. WANG, K. (1984). In: *Contractile mechanisms in muscle* (Pollack, G. H. and Sugi, H. (Eds)), Plenum Press, New York, p. 285.
130. MARUYAMA, K. and KIMURA, S. (1985). In: *Cell motility: mechanism and regulation* (Ishikawa, H., Hatano, S. and Sato, H. (Eds)), University of Tokyo Press, Tokyo, in press.
131. LOCKER, R. H. (1984). *Food Microstructure*, **3**, 17.

Chapter 3

PRE-SLAUGHTER STUNNING

T. M. LEACH

School of Veterinary Studies, Murdoch University, Murdoch, Western Australia

1. INTRODUCTION

In those situations where considerations of animal welfare play little part in the regulation of human activities, it is the usual practice to slaughter calves, pigs, sheep and goats by restraining them manually while the major blood vessels of the neck are severed and the animals bleed to death. Large calves[1] and pigs may receive a blow on the head to incapacitate them before the neck is incised. The greater strength of adult cattle makes more forceful restraint essential. It has been the general practice to immobilise them by a blow on the head with a felling hammer or club, before killing them by exsanguination. In some Mediterranean, Latin American and South East Asian countries, the animal's head is pulled close to a ring in the floor or wall and a broad bladed, double edged knife (*puntilla*) is plunged into the altanto-occipital space, severing the spinal cord and producing immediate quadriplegia.[2,3] The animal is then bled out. Another method, used by some Nilotic cattle-owning tribes, is to kill the animal by a direct stab into the heart through the chest wall. Poultry are commonly killed by decapitation or by dislocating the neck, although some are suspended by the legs and the vessels of the neck are then severed.

By the early eighteenth century a mediaeval weapon, the poleaxe, had been taken into use for stunning cattle.[4] One side of the head was an axe and the other was a hammer. Subsequently, the hammer was modified into a cylindrical rod about 4 cm long and 1 cm diameter. When accurately aimed, the rod penetrated the skull and entered the brain. In the latter part of the nineteenth century, various novel stunning devices were introduced, including a mask that incorporated a bolt. The mask was secured to the animal's head so that the bolt was located over the frontal area and penetrated the brain when struck with a hammer. A logical development was the introduction of the captive bolt pistol at the beginning of the twentieth century. Rasenack [5] has summarised the historical development of slaughtering methods and Fahrbach [6] reviewed the more recent changes up to the introduction of the pistols currently in use.

2. LEGAL REQUIREMENTS

During the early 1900s there was increasing pressure from animal welfare organisations for the mandatory stunning of slaughter stock before they were bled out. As a result, this procedure became a statutory requirement in many countries by the middle of the present century. Within Great Britain, the practice was first enforced by the Slaughter of Animals (Scotland) Act, 1928, and the Slaughter of Animals Act, 1933, which applied in England and Wales.

Shooting, either with a free bullet or captive bolt pistol, and electrical stunning were the generally accepted methods. The use of electrical stunning was authorised in the 1933 Act by the statement that a mechanically operated instrument included an instrument for stunning by means of electricity and was specifically approved in the Slaughter of Animals Act, 1958, which repealed the 1933 Act. The new legislation required any animal to which the Act referred, unless it was slaughtered instantaneously, i.e. shot with a free bullet, to be stunned, either by means of a mechanically-operated instrument, e.g. a captive bolt pistol, or by an instrument for stunning by means of electricity so that it was instantaneously rendered insensible to pain until death supervened. The Act empowered the responsible Minister to approve other methods of stunning, provided that the animals were 'rendered insensible to pain until death supervenes'. Thus the pre-slaughter immobilisation of pigs with carbon dioxide was authorised by the Slaughter of Pigs (Anaesthesia) Regulations, 1958. These stipulated that the gas concentration should be maintained between 62% and 70% and included various requirements in respect of warning devices if gas concentrations went outside these limits, as well as arrangements providing access to pigs in the plant if a breakdown occurred. The 1958 Act has since been replaced by the Slaughterhouse Act, 1974,[7] but these provisions remain unchanged.

The legislation in force in the European Economic Community during the mid 1970s was collated by von Mickwitz and Leach.[8] All the member States required animals to be stunned before slaughter. In addition to stipulating approved stunning methods, some European legislation contains detailed requirements for their application, e.g. the Dutch requirements on electrical stunning,[9] while other countries rely on advisory literature, e.g. the United Kingdom's official circulars on electrical stunning,[10,11] or on informal bulletins.[12] All developed countries have similar legislation.

The impact of existing laws on stunning practice is commented on in the appropriate sections below. Everton[13] has discussed the need for reform of the British legislation, suggesting that, although full of good intentions, it has failed to protect animals adequately. His proposals are in general agreement with those of von Mickwitz and Leach,[8] and are discussed in Section 7.

3. HANDLING BEFORE SLAUGHTER

In addition to welfare considerations, handling of animals with the minimum of disturbance up to the point of slaughter is important in relation to meat quality[14] and Lister *et al.*[15] have detailed the varied responses of slaughter stock in this situation. Aspects of worker safety and efficiency also demand attention. All these factors should influence the design of slaughtering facilities and Grandin[16,17] has dealt with this topic in some detail.

In principle, stock should be brought to the point of slaughter as quietly as possible and the application of knowledge gained from studies on animal behaviour is playing an increasing part in improving plant layout. Detailed consideration of lairage design is beyond the scope of this chapter, but it is not possible to ensure that a steady flow of animals is maintained by direct release from individual pens. Provision must be made for stock to be fed easily into a race system incorporating a crowding pen; this ensures a reservoir of animals awaiting slaughter. A second race, of adequate length, should lead from the crowding pen to the slaughter point. Throughout the system walls should be smooth, solid and sufficiently high to prevent animals from being distracted by seeing out. Floors should be non-slip and races should be curved, so that animals see those in front disappearing round the bend and will tend to follow. Where it is necessary to insert gates across races, they should be of open construction to ensure that animals remain in visual contact with those ahead. Footways for operatives should be provided inside the inner curvature of races.

Facilities at the point of slaughter will differ with the class of stock being handled and the size of the plant. Adult cattle, particularly if extensively reared, are usually confined in a stunning box, which is a legislative requirement in some countries.[8] After stunning, a horizontally hinged flap on the side of the box is released and the animal is ejected. It is then hoisted by one or both legs, stuck and moved to the bleeding passage. In some plants cattle are stuck before hoisting, but the procedure is otherwise the same. Stunning boxes are strong and durable; however, apart from their ability to contain the animal, little thought has been given in their design to the behaviour of cattle in confined spaces or to worker safety and efficiency. In small to medium sized abattoirs groups of sheep and pigs are generally stunned in a pen, while plants with higher throughputs usually feed the stock to the stunning point by a restraining conveyor, in which animals are carried forward between moving belts. Restraining conveyors are also used for cattle in some of the major beef works. These systems will be considered in Section 7.

4. ASSESSMENT OF THE ONSET AND DURATION OF INSENSIBILITY

The introduction of legislation intended to protect the welfare of animals at slaughter was itself a recognition of the fact that the empirical views of slaughtermen on the efficiency of stunning methods, based on whether their task was facilitated, did not provide an adequate assurance of insensibility. Blackmore[18] has defined the primary objective with the statement 'First, and over-riding all other considerations and to meet ethical and statutory requirements, the process must be humane'.

Legislation commonly requires stunning to be 'instantaneously effective'. In practice this might be interpreted as within one second and such a definition would probably command wide acceptance. To decide if a method of stunning meets this criterion is, of course, dependent on the capacity to determine whether a state of insensibility exists after its application. Apart from the inherent complexity of reliable tests differentiating sensibility from insensibility, operating conditions in slaughterhouses tend to make their application difficult, if not impractical. Consequently, assessments of the efficiency of stunning methods in commercial practice are likely to be subjective and so not susceptible to critical analysis.

In these circumstances, judgement has to be based on the extrapolation of experimental findings. Discussing this difficulty, Blackmore and Newhook[3] pointed out that interpretation of the data, especially in relation to pain, is necessarily largely subjective and sometimes anthropomorphic. In considering the latter, they quote Kitchell[19] 'Since the ascription of pain perception to animals is only feasible by reasoning from analogy, a knowledge about pain in animals remains to be a *probable inference*. Sheer neglect of the probabilistic nature of pain perception in animals leads to total anthropomorphism. On the other hand an overemphasis of the uncertainty of our knowledge about pain perception in animals leads to a denial that pain perception exists in animals, which is logically as well as empirically unfounded'.

Observation of the presence or absence of various reflex responses after pre-slaughter stunning has been advocated as a basis for assessing its efficiency. However, as Blackmore and Newhook[3] comment, excepting carbon dioxide anaesthesia, all the methods of pre-slaughter stunning in use cause insensibility in an entirely different manner from chemical anaesthesia. Consequently it may not be appropriate to assess them on the same basis when judging insensibility as when gauging the depth of anaesthesia. The palpebral and corneal reflexes are probably most commonly used for this purpose. They are initiated by excitation of the exposed surface of the eye or of the conjunctiva and are usually elicited by touching, or attempting to touch, these structures. In chemical anaesthesia narcosis of the cortex and of its connections with subcortical centres precedes that of the paths of these reflexes. The corneal reflex may persist to the stage of deep anaesthesia or almost until death[20] and it follows that, if the reflex is not functioning, it can be assumed in the anaesthetised animal that the cortex has already become inactive. However, in the case of electrical stunning, Roos and Koopmans[21,22] showed that passage of alternating current through the central nervous system causes tonic spasm of the eyelid muscles, inhibiting these reflexes, which can then usually only be elicited again some 40–90 s after current flow has ceased. On the basis of their findings, these reflexes cannot be regarded as reliable guides to the efficiency of electrical stunning.

In chemical anaesthesia the pupillary light reflex is elicited by closing the eyelids over both eyes for ten to twenty seconds, after which one eye is allowed to open and a strong light is shone into it. If the reflex is present, the exposed pupil will contract. According to Hall,[23] the reflex is abolished about half way through the third plane of surgical anaesthesia. Croft[24] stated that the pupillary reaction to light does not involve any of the higher regions of the brain; it merely indicates that the brain is functioning at the level of the Edinger–Westphal nucleus. In observing rabbits recovering from electrical stunning, she noted that this reflex reappeared first, closely followed by the recovery of the cardiac pain reflex, *see below*. Since the reflex depends only on the lower parts of the brain functioning, its abolition would appear to be a reliable indication of insensibility. It has nevertheless received comparatively little attention in published work on pre-slaughter stunning techniques.

Experiments by Croft[25] established the value of the cardiac pain reflex as a means of sensibility assessment in the laboratory. A painful stimulus caused the heart rate of a rabbit to increase by 20–30 % for about 15 s. The magnitude of the increase was not greatly affected by the strength of the stimulus. Deep anaesthesia or electrical stunning abolished the reflex, which subsequently returned and this could be used as an indication of pain perception. Moreover, since no voluntary muscles were involved, it could be applied in animals treated with muscle relaxant drugs or to monitor for the occurrence of electrical curarisation, Croft's second stage of recovery in electrical pre-slaughtering stunning. In the course of these experiments she also showed that the pedal, or paw retraction, reflex was still inhibited after the return of the cardiac pain reflex, indicating that, although the animal's sensibility had returned, it was incapable of voluntary muscular movement. Croft[24] considered that the pattern of physical response of animals to the passage of electric current through the brain could be used in judging the effectiveness of stunning. As this approach is confined to the assessment of electrical pre-slaughter stunning, it is discussed in Section 5.2.1.

The electroencephalogram (EEG) or electrocorticogram (ECoG) provides a means of objectively assessing the onset and duration of insensibility following pre-slaughter stunning in the laboratory. Lopes da Silva[26] has discussed recent developments in the application of these techniques in this field. According to Ruckenbusch,[27] cortical activity in the fully sensible adult ruminant is characterised by rapid waves of small amplitude (10–30 Hz, 20–50 μV). In the resting state, frequency diminishes to below 10 Hz and amplitude rises to about 50 μV. Sleep is characterised by very low frequency (6 Hz) and strongly increased amplitude (100–150 μV). It is widely accepted (for example, Fricker and Riek;[28] Lambooy and Spanjaard[29]) that, when δ (<4 Hz) and θ (4–8 Hz) waves occur immediately after captive bolt stunning, it is almost certain that sensibility has been lost and that this is definitely the case when the trace becomes iso-electric. An alternative basis for interpreting EEG traces in sheep during experiments on slaughtering methods was adopted by

Newhook and Blackmore,[30] who determined the limits of amplitude they considered covered the range of sensibility in sheep. Their bottom limit, 10 μV, was the lowest level observed in somnolent sheep, the top limit, 35 μV, was higher than any recording from highly excited animals. They assumed that within these limits sheep might be sensible. Gregory *et al.*[31] used the EEG and ECoG to study visually evoked cortical potentials in anaesthetised sheep, in the development of methods of assessing sensibility and brain death during stunning and slaughtering. This technique has an advantage over monitoring spontaneous brain signals, which involves the assumption that certain types of signal indicate a loss of brain responsiveness, in that the evoked response provides a measure of the brain's reaction to a given stimulus. It was found in this species that the maximum displacement of visually evoked potentials was 7·5 times greater with the ECoG than with the EEG.

5. PRE-SLAUGHTER STUNNING METHODS

5.1. Firearms

5.1.1. Free Bullet

Although the use of conventional firearms for slaughtering animals does not fall within the category of pre-slaughter stunning, it provides an effective means of killing them for meat production. The hazards associated with using a free bullet, particularly in a confined space such as an abattoir, need not be emphasised. Nevertheless, there are circumstances in which a strong case can be made for its use. It is a commonly held view amongst those experienced in killing horses that captive bolt pistols are unsatisfactory for this purpose[32] and this species is normally shot with a free bullet. Massive development of the cranial bones and overlying tissues in the frontal region of mature bulls prevents the standard captive bolt pistols from functioning adequately. Although heavier models are available for use solely with bulls, in most circumstances a free bullet is the only readily available, effective means of slaughter. It is unsafe to use a captive bolt pistol for stunning adult boars and sows, Section 5.1.2.3, and electrical stunning is not satisfactory for these mature animals, Section 5.2.6, so that shooting with a free bullet is the method of choice. A variety of specialised instruments is available, firing bullets of various calibres; in the United Kingdom the most usual are 0·22, 0·31 and 0·32 in. (5·59, 7·87 and 8·13 mm, respectively). A pistol with a spring loaded barrel, that can only fire when the muzzle is pressed against the head, is an effective and safer instrument than a conventional firearm.

5.1.2. Captive Bolt

Reference has already been made to the development about the turn of the century of captive bolt pistols, to replace the poleaxe. In a study of head wounding, Butler *et al.*[33] found that the passage of a high velocity missile produced a large temporary cavity within the brain. The resulting injuries were attributable to the great increase of internal pressure this caused. Hopkinson and Marshall[34] also concluded that, in addition to their lacerating affect, high velocity missiles caused injury by producing a temporary pulsating cavity within the tissues, the magnitude of which depended on the missile's rate of retardation. Lambooy and Spanjaard[29] distinguished three ways in which a missile could injure the brain; laceration and crushing at low velocity, shock wave damage at 100–300 m/s and temporary cavitation at higher velocities. They stressed that the degree of injury was primarily dependent on the energy transmitted to the brain. Lambooy[35] studied the force necessary for a captive bolt pistol to penetrate heads from young bulls (*c.* 18 months), veal calves (*c.* 6 months) and baconers (*c.* 100 kg live weight). Five heads were used from each class of animal and five points within 2 cm radius of the recommended frontal stunning sites for both species[12] were tested on each head; no significant intragroup differences were found. The average maximum force to achieve penetration was $30{\cdot}6 \pm 6{\cdot}5\ kg/mm^2$ for young bulls, $23{\cdot}9 \pm 5{\cdot}4\ kg/mm^2$ for veal calves and $32{\cdot}7 \pm 8{\cdot}8\ kg/mm^2$ for baconers. On the assumption that the maximum forces needed to penetrate the heads were normally distributed, stunning would be effective in 99·8 % of these animals if $50{\cdot}1\ kg/mm^2$ was used for young bulls, $40{\cdot}1\ kg/mm^2$ for veal calves and $59{\cdot}1\ kg/mm^2$ for baconers, based on the average values plus three times the standard deviation.

By measuring the depth of penetration of the bolt into lead cylinders, Lambooy *et al.*[36] compared the forces generated by cartridges of different charges supplied for use with three commercially available captive bolt pistols. The summarised results are shown in Table 1.

On the basis of these results, the cartridges tested yielded the levels of force required to stun the classes of cattle for which they were designated, excepting the red Cash cartridge, which was inadequate for young bulls. None of the cartridges tested attained the estimated requirement for reliable stunning of bacon weight pigs. In considering these results, it should be remembered that they were obtained with pistols and cartridges

TABLE 1

THE MAXIMUM FORCE DEVELOPED BY CARTRIDGES OF VARIOUS STRENGTHS (AS INDICATED BY DIFFERENT COLOURS) SUPPLIED BY CASH,[a] KERNER[b] AND SCHERMER[c] CAPTIVE BOLT PISTOLS

	Cash		*Kerner*		*Schermer*	
	Colour	*kg/mm²*	*Colour*	*kg/mm²*	*Colour*	*kg/mm²*
Light weight animals	pink	47·2			green	52·7
			yellow	50·2	yellow	54·4
Heavy weight animals	red	49·1	red	53·1	red	55·5
	black	51·7			black	56·0

[a] Cash Allrounder, Accles & Shelvoke Ltd, Birmingham, United Kingdom.
[b] Kerner, Karlheim Kerner, Leverkusen-Rheindorf, Federal German Republic.
[c] Model ME/MS, Karl Schermer & Co., Karlsruhe, Federal German Republic.
(from Reference 36, by permission of Martinus Nijhoff, The Hague).

operating in optimum conditions. Lack of maintenance and wear and tear will reduce efficiency.

5.1.2.1. Cattle. Captive bolt pistols are the most commonly used instruments for pre-slaughter stunning of cattle of all ages. Using captive bolt stunning in the frontal region, Gross[37] studied the ECoG of five calves, aged 3 to 4 weeks and weighing between 40 and 60 kg. Immediately after shooting, there was extreme general disturbance of the trace with waves of 1–2 Hz predominating. He concluded that this reaction almost certainly excluded the perception of pain. In four calves the trace became iso-electric within 28 s of shooting, in the fifth after 73 s.

Lambooy and Spanjaard[29] compared the effect of stunning 6 month old calves, weighing about 200 kg, with a captive bolt pistol at three different sites, frontally, occipitally and in the nape of the neck. δ and θ waves appeared on the ECoG immediately after shooting frontally or at the occiput and the traces subsequently became iso-electric. Although occipital stunning did not cause macroscopic damage to the cortex, as did frontal stunning, the authors concluded that shooting at either site caused immediate insensibility. Calves shot in the nape of the neck continued to show a normal ECoG for an average of 21 s (SD $\pm$ 6 s) before it was judged from changes in the trace that they were insensible. This finding was in agreement with subjective observations on shooting at this site, during the survey of pre-slaughter stunning in the European Economic Community.[8]

In a study by Fricker and Reik[28] on twelve cows shot in the frontal area with a captive bolt pistol, they noted similar ECoG changes on shooting to those reported in calves shot at the same site, except that the trace immediately became iso-electric in two cows, changing to the slow, large amplitude wave pattern after 10–30 s. In eleven of the cows the trace was iso-electric within 120 s. Although the trace from the twelfth animal did not become iso-electric until 30 s after sticking, which took place over 6 min after shooting, the authors concluded, from the instantaneous appearance of evidence of brain stem lesions in all the animals, that they could not have experienced any pain.

In cattle the brain is nearest to the surface of the head in the frontal area and this is the site of choice for captive bolt stunning. There is a dearth of experimental evidence on the effectiveness of pre-slaughter stunning of adult cattle in other sites. Major structural changes occur in the bovine skull during growth, particularly in the size and extent of the sinuses. This suggests that stunning older cattle elsewhere than in the frontal area may be less effective than in calves.

Pneumatically powered captive bolt stunners are used in some plants slaughtering large numbers of cattle. Although their muzzle velocity is lower than that of cartridge operated pistols, they have bolts of greater diameter and length, capable of causing extensive damage to the brain. No critical studies appear to have been published on their effect on brain function.

5.1.2.2. Sheep. Freesemann[38] studied the ECoG in ten sheep, weighing 30–40 kg stunned with a captive bolt pistol in the frontal area. Shooting caused an immediate, general disturbance of the trace, with δ waves predominating. In eight animals the trace became iso-electric within 17 s of shooting. He concluded that the sheep were almost certainly insensible to pain when slow, large amplitude waves were present and definitely were when the trace became iso-electric.

By means of a detached bolt, modelled on that of a commercial captive bolt, Lambooy[39] studied the effect of bolt velocity on the ECoG of sheep weighing about 45 kg. Sixteen sheep had their skulls trephined in the frontal area and the detached bolt was driven into the brain through this hole by a tap with a hammer. Eight similar intact sheep were stunned at the corresponding site with a pistol having a bolt of the same dimensions. Depth of penetration with both instruments was controlled by plastic rings round the bolts. Six sheep were stunned to a depth of 6 cm with the detached bolt. The ECoG immediately showed δ (<4 Hz) and θ (4–8 Hz) waves. When their brains were examined, the cortex and deeper parts of the brain had been damaged by the bolt. Six sheep stunned to a depth of 3 cm appeared unstable, but were able to stand. In addition to α (8–13 Hz) and β

(>13 Hz) waves, the ECoG also showed δ waves. On examination, the cortex was damaged, but the ventricle was intact.

Shooting to a depth of 2·2 cm with the captive bolt pistol resulted in the ECoGs of four intact sheep immediately altering to δ and θ waves. On post-mortem, three sheep showed damage to the cortex only; a bone fracture had resulted in damage to deeper parts of the brain of the fourth animal. Shooting four intact sheep to a depth of 0·5 cm, i.e. just beneath the skull, resulted in the trace of one sheep being unchanged. At subsequent examination, the cortex appeared to be slightly damaged. The second and third sheep had δ waves in their ECoGs, in addition to α and β waves, indicating a state of concussion. The cortex was damaged to a depth of about 2 mm in these two animals. After showing a similar trace to that of the second and third sheep for 21 s, the ECoG of the fourth sheep changed to δ and θ waves only. Post-mortem examination showed that a bone fracture had damaged the cortex.

Four trephined sheep were shot with the captive bolt pistol without penetration of the brain and their ECoGs remained unchanged. At autopsy, one animal had haemorrhaged below the dura mater and this was attributed to increased air pressure over the brain at the time of shooting. The author concluded that, if the captive bolt did not enter the cranium, insensibility did not occur because insufficient energy was transmitted to the brain. To be effective, a low velocity bolt needed to be sufficiently long to lacerate and crush the deeper parts of the brain. The velocity of a captive bolt pistol penetrating to 2·2 cm produced an adequate shock wave to cause insensibility. To achieve a satisfactory stun, a captive bolt pistol must meet one of these criteria.

Schütt-Abraham *et al.*[40] monitored extradural pressure changes in sheep, weighing between 50 and 100 kg, stunned at the occiput either with a captive bolt or a concussion stunner of equivalent velocity. They obtained peak values of 9–30 bar (mean value 23) with the former and 3–16 (mean value 11) with the latter. Captive bolt shooting of 11 sheep caused immediate extinction of the corneal reflex in all cases and respiration stopped at once in nine. They concluded from these results that captive bolt stunning was apparently adequate. Captive bolt velocities measured during these observations ranged between 28·7 and 57 m/s, which implied that the injuries they caused would be due to laceration and crushing.[29] The concussion stunner results are discussed in Section 5.1.3.2. These workers also drew attention to the low standard of stunning prevalent in practice; of 166 heads examined from sheep slaughtered in an abattoir using captive bolt stunning, only 97 had been shot at the recommended site, 47 showed either badly sited shots or none at all, while 22 showed evidence of blows to their heads.

Because of the location of the cranium within the sheep's skull, the recommended site for captive bolt stunning is at the occiput. Stunning can be carried out in the frontal area, but the presence of sinuses is liable to reduce the reliability of shooting at this site. In addition, sheep are more likely to shy away if approached from the front.

5.1.2.3. Pigs. The configuration of the pig's skull makes captive bolt stunning unreliable for all but light animals and reference has been made to the inability of standard captive bolt pistols to stun 100 kg pigs consistently. In adult pigs the sinuses have reached full development and the skull bones are dense. The captive bolt is likely not only to fail to penetrate the cranium, but may jam in the overlying bone, precipitating a dangerous situation. Large pigs should either be shot with a free bullet or stunned with an adequate electric current, Section 5.2.6. The practical difficulties of stunning adult pigs have been clearly described by Parker.[41]

5.1.3. Concussion Stunning

In considering the causation of cerebral concussion, Holbourn[42] postulated that sudden impact resulted in rotational deformation of the brain, some parts moving greater distances than others in relation to fixed points on the cranium. This led to shearing forces damaging the brain tissues and loss of consciousness. However, Lambooy and Eikelenboom[43] defined the condition as a traumatically induced derangement of the nervous system, resulting in instantaneous diminution or loss of consciousness without gross anatomical changes in the brain. On the basis of work by Walker *et al.*[44] and Nilsson and Pontén,[45] they concluded that the most important factor was the rapid oscillation in intracranial pressure, rather than the actual pressure developed by the waves resulting from a blow on the head.

The introduction of an instrument capable of delivering a controlled blow to the head was a logical means of replacing stunning with a hammer or poleaxe. The majority of concussion stunners work on the same principle as a captive bolt pistol, with a blank cartridge as the propellant, although pneumatically powered models are also available. The captive bolt is replaced by a rod terminating in a non-penetrating, convex (mushroom shaped) extremity that is driven against the surface of the head.

Although at present accepted for pre-slaughter stunning under the legislation of many countries, concussion stunners are less widely used than captive bolt pistols. They were banned in The Netherlands in 1920

(Lambooy, 1983; personal communication) because of doubt about their ability to stun effectively. There is a dearth of publications on experimental work with percussion stunners in adult cattle and pigs.

5.1.3.1. Calves. Using a pneumatic captive bolt stunner,* modified to deliver a concussive blow and operating at 35–50 kg/cm^2 air pressure, Blackmore[46] stunned 90 1–2 week old calves at the poll (occipitally) and concluded that 80% were satisfactorily stunned, on the basis of immediate collapse with signs of extensor rigidity and some generalised muscular tremors. The corneal reflex often returned within 20 s, but there was no sign of a righting reflex for at least 60 s. However, the brains of the calves judged to have been successfully stunned all had areas of gross haemorrhage, making them unacceptable for human consumption.

Lambooy *et al.*[47] studied the ECoGs of 6 month old calves, weighing about 200 kg, stunned with a commercially available stunner.† Using cartridges suitable for lightweight cattle, they obtained traces from 19 calves stunned frontally and from 8 stunned at the poll. In 15 of the first group and 3 from the second, stunning was accompanied by δ and θ waves, the trace becoming iso-electric after an average of 32 s (SD ± 21 s). Of the other 9 calves, 7 were not rendered insensible and 2 showed a reduction in wave frequency for about 2 min, but were judged from the character of the trace not to have lost sensibility. The authors suggested that the low level of success in calves stunned at the poll may have been influenced by the automatic (contact) trigger striking soft tissues in the occipital region and failing to ignite the cartridge.

In contrast to Blackmore's findings,[46] only half the calves showed cerebral haemorrhage after stunning. Five animals had fissures in the skull and 20 had fractures. On the basis of Blackmore's and their own findings, the authors concluded that concussion stunning of calves was insufficiently effective to be an acceptable method of pre-slaughter stunning.

5.1.3.2. Sheep. A cartridge operated captive bolt stunner‡ was modified by Blackmore[46] for concussion stunning of lambs. Nineteen of twenty lambs were effectively stunned at the poll with a $1\frac{1}{4}$ grains (81 mg) cartridge, judged from their immediate collapse with signs of extensor rigidity and some generalised muscular tremors. Signs of recovery and the development of righting reflexes did not usually occur in less than 2 min; during this period the corneal and palpebral reflexes were absent. Brains from six of these lambs showed haemorrhages. Lowering the impact velocity of the

* Hantover Inc., Kansas City, Missouri, United States of America.
† Cash Knocker. Accles and Shelvoke Ltd, Birmingham, UK.
‡ Cash Accles and Shelvoke Ltd, Birmingham, UK.

stunner, by reducing its travelling distance, reduced the incidence of brain haemorrhage, but also resulted in fewer lambs being stunned.

Attempts to stun adult sheep with this instrument, using a 2 grains (130 mg) charge, were unsuccessful. However, the modified pneumatic stunner employed to stun calves in the experiment reported above (Section 5.1.3.1), effectively stunned small groups of lambs, hoggets and ewes, operating at pressure between 35 and 100 kg/cm^2. The high impulses necessary to stun adult sheep inevitably caused brain haemorrhages. By discharging these stunners at a suspended steel weight, the author calculated the impulse generated from measurements of the pendulum's lift and travel.

Reference has been made above (Section 5.1.2.2) to an investigation by Schütt-Abraham *et al.*[40] which compared the extradural pressure changes occurring in sheep stunned at the poll with a captive bolt or a concussion stunner. Although the bolt velocities were similar (means 52 m/s and 49 m/s, respectively), the peak extradural pressure generated by the captive bolt stunner was double that produced by the concussion stunner (means 23 bar and 11 bar). The concussion stunner generally caused fissures and fractures radiating from the point of impact and the parietal bone was extensively damaged in two cases. Respiration ceased immediately in both these sheep and the corneal reflex was abolished; in the other four sheep in this group, respiration ceased for 7–43 s and then resumed, while the corneal reflex was maintained in three and returned after about a minute in the fourth. The authors concluded that, under the conditions of this experiment, the results of concussion stunning were not satisfactory.

5.2. Electricity

There were sporadic reports from the latter part of the eighteenth century onwards about the use of electricity to induce anaesthesia in animals, including man, as well as on its possible application to pre-slaughter stunning. Warrington[48] has reviewed this historical literature and also subsequent developments up to the early 1970s.

The critical factor in achieving an effective stun with electricity is the level of current flowing through the brain; this is determined by the voltage, the highly variable resistance of the animal's tissues and by the positioning of the electrodes. Although improvements in stunning have been reported with modified wave forms[49] and higher frequencies,[50,51] recent research has tended to concentrate on increasing the stunning efficiency of 50 or 60 Hz mains alternating current by raising the voltage, and hence, the amperage. Working with pigs, Hoenderken[52] found that it was necessary to apply 300 V, giving a current flow exceeding 2 A, to achieve consistently effective stunning; Blackmore and Peterson[53] reported satisfactory stunning of lambs with 400 V and a current flow of *c.* 0·8 A.

5.2.1. Assessment of Insensibility After Electrical Stunning

Assessment of insensibility is of particular importance in electrical stunning because of the possible occurrence of paralysis of the voluntary muscles, without loss of sensibility to pain, the nightmare state described by Hume.[54] Using ECoG recordings in rabbits, Croft[24] studied the electroplectic fit that is precipitated by successful electrical stunning. On the basis of the cardiac pain reflex (Section 4), she found that pain perception returned about 40 s after effective electrical stunning, but was accompanied by a phase of electrical curarisation lasting approximately 30 s. However, if two electrical stimuli were given in quick succession, the state of muscular paralysis was prolonged for about 10 min.

In the course of this study, Croft examined the response to electrical stunning of various species and found a high degree of consistency between the behaviour of rats, rabbits, pigs, sheep and man. A minimum current of 25 to 30 mA was necessary to produce insensibility in rabbits and alteration of other factors would not compensate for inadequate current flow, nor would increasing the time of application compensate for inadequate voltage. Frequencies between 50 and 200 Hz were satisfactory; 25 Hz caused the muscles to twitch synchronously with the pulse. Croft stated that frequencies of 500 Hz or above had been found painful in human experience. Using a minimum pulse width of 2 ms, current flow for 2 s was adequate to produce insensibility and increasing either of these factors made little difference to the result. These observations were confirmed in sheep and, later, in pigs subjected to electrical stunning in slaughterhouses.

In defining the sequence of events when stunning successfully evoked an electroplectic fit, Croft stressed that the critical feature was extensor tonus of the hind limbs during the first stage. When this occurred, the animal was effectively stunned, judged on the basis of a flat ECoG trace, indicating the cortex was inactive. This stage also included rigid extension of the forelimbs, opisthotonus and arrest of respiration. After 10 s the limb muscles gradually relaxed and clonic contractions occurred. About 20 s later respiration resumed and the animal became flaccid, but lay motionless for about 3 min. Current flow might precipitate any of these signs, but, unless extensor tonus occurred in the hind limbs, the animal could not be regarded as effectively stunned. Although these criteria were established over thirty years ago and are often difficult to interpret in the slaughterhouse, when present they remain the only reliable means of assessing pre-slaughter electrical stunning without recourse to EEG examination.

In an attempt to determine whether the initiation of electrical stunning was painful, Leach *et al.*[55] subjected groups of five sheep to 90 V, 50 Hz for 3, 6, 9, 12 and 15 s. Over an 11 day conditioning period, the sheep were exposed daily to a light for 5 s immediately before current flow. On the twelfth day the sheep were exposed to the light, but not to the current. Heart rate, packed cell volume and plasma glucose levels were monitored to assess the animals' physiological response. The light did not affect the heart rate, but both packed cell volume and plasma glucose levels rose sharply after stunning immediately following exposure to the light. However, neither value rose on the twelfth day, when the sheep were not stunned. It was concluded that, under the conditions of this experiment, the initiation of electrical stunning was not painful, as it failed to elicit a conditioned response. The depth and duration of insensibility following electrical stunning was not investigated.

The observed response to electrical stunning may be modified by the duration of current flow. In studying the application to pigs of voltage ranging from 70 V to 380 V, 50 Hz, Müller and Engeli[56] varied the exposure time from 1 s to 18 s. They termed brief exposure, from 1 s upwards, *electro-shock* and exposure from 14 s upwards *electro-coma.* They considered either practice gave a satisfactory stun, if 70 to 140 V was used, but brief exposure precipitated violent convulsions that made hoisting and sticking difficult. The less violent response in *electro-coma* facilitated slaughtering.

An important factor in determining the acceptability of electrical stunning is the need to ensure that exsanguination takes place sufficiently quickly to prevent the return of sensibility. In a recent survey of abattoirs slaughtering sheep in Great Britain, Gregory *et al.*[31] found that the average interval between the end of stunning and sticking was 23 s. Laboratory studies[57] had shown that the time between the severance of both carotid arteries and jugular veins and the loss of brain responsiveness, judged by visually evoked cortical responses, was 14 s. To achieve satisfactory stunning in these circumstances, insensibility must therefore last at least 37 s. However, the interval between stunning and sticking was considerably longer than 23 s in some abattoirs and it is well established that other sticking techniques, e.g. severing the carotid artery and jugular vein on one side of the neck or cutting the jugular veins only, considerably delay the loss of sensibility, judged either by the appearance of the ECoG trace[30] or the loss of visually evoked responses.[31] Apart from failure to apply adequate current to achieve a satisfactory stun, which is often due to faulty application of the electrodes, delay between stunning and sticking is likely to be the most serious defect in electrical stunning practice at present.

5.2.2. Cardiac Function and Bleeding Out

Probably the major development in electrical stunning during the past decade has been the demonstration that heart action can be stopped during stunning of pigs, sheep and poultry without affecting the quality of the carcass. Killing animals in this way also eliminates the problem, discussed in the preceding section, of their recovery after electrical stunning because of delayed exsanguination. It has been a traditional belief in the meat industry that heart action is necessary to ensure satisfactory bleeding out of the carcass, although consideration of hydrostatic principles suggests that, once the major vessels have been severed, return flow of blood falls dramatically and heart action will then play little part in effecting bleed out.

In reviewing data on the residual blood content of meat, Warriss[58] commented that there was confusion between the amount of blood lost at slaughter and the quantity remaining in the carcass musculature. Thus it had been shown[59] that more blood was collected from sheep bled horizontally than those bled vertically and that electrically stunned sheep yielded more blood than those stunned with a captive bolt. Nevertheless there was no significant difference in the residual haemoglobin levels in the *M. longissimus dorsi* of the different groups; nor was it related to the amount of blood lost at slaughter. When a discrepancy occurs, it seems likely to be due to differences in the quantity of blood trapped in the viscera and major blood vessels. Warriss and Wotton[60] compared groups of electrically stunned pigs, one of which was bled out normally and the other had the heart put into ventricular fibrillation before bleeding out. Cardiac arrest failed to affect the weight or rate of blood loss and there were no significant differences between groups in the total haem pigment concentration in either *M. longissimus dorsi* or *M. semimembranosus*. Objective colour measurements of the meat failed to reveal any differences in appearance. In observations on pigs stunned with current rising to 1·1 A, Swatland[61] found there was rarely any evidence of ventricular contraction during the maximum rate of exsanguination and concluded that pigs could be effectively bled out when stunning had stopped the heart. Similar findings in the stunning of lambs[62] and poultry[63] are described in Sections 5.2.5 and 5.2.7, respectively.

5.2.3. Adult Cattle

An account of early investigations on the stunning of adult cattle with electricity is included in Warrington's review.[48] Despite some favourable reports, the technique did not gain widespread acceptance, possibly because it had not been adapted to meet the operational requirements of plants with large daily throughputs. A notable exception was the Mikoyan slaughterhouse in Moscow.[64] Four cattle could be accommodated in each of four boxes. These had metal floors connected to a 200 V, 50 Hz supply. The current, *c.* 2·5 A, was switched on for 10–30 s. An operative applied an electrode carrying the same current to the heads of any animals not falling down immediately. The cattle were ejected automatically and were then shackled and hoisted onto four separate dressing lines. The plant could handle 360 head per hour. Cattle did not move during bleeding out (Hoenderken, 1976; personal communication), having apparently been killed by electrocution. A similar stunning system was in use for pigs.

5.2.4. Calves

The effect of electrode position on the stunning in slaughterhouses of calves about a week old was studied by Blackmore and Petersen.[53] Three groups of calves all had two electrodes applied to the head, the second group also had an electrode applied to the dorsal surface of the mid-thoracic region (head-to-back) and the legs of the third group were in contact with strip electrodes (head-to-leg). The stunner was connected to a 50 Hz supply, giving up to 400 V and the current was variable between 0·5 and 2 A. Normal cardiac function was detected in only one of 250 calves stunned with *c.* 1 A for 5 s by the head-to-back equipment. In another trial, no normal cardiac function was detected in 100 calves stunned with 2 A for 5 s by the head-to-leg method. Reducing the current to 0·8 A still resulted in at least half the calves developing cardiac dysfunction, there being good agreement between examination by ECG or auscultation. No differences were detected in the corneal and palpebral reflexes of calves with or without heart function at the time of slaughter. The authors concluded that it was impractical to use these reflexes or such responses as respiratory gasps, vocalisation or kicking, as indications of insensibility in calves. They referred to the finding of Newhook and Blackmore[65] that, unless cardiac dysfunction accompanied efficient stunning or exsanguination began within 20 s of stunning, sensibility was likely to return before death supervened, because the supply of blood to the brain in this species by the vertebral artery was not immediately interrupted by severance of the carotid arteries.

The electrical pre-slaughter stunning of larger calves does not appear to be widely practised.[8] In a study of electrical stunning of calves weighing 200 kg, Lambooy and Spanjaard[66] estimated that a current of 0·87 A was necessary to stun 90% of the animals effectively. They used scissor type tongs and found that atrial fibrillation often occurred when the tongs were applied to the ear and opposite eye. They concluded that the relatively short electroplectic fit and the clonic spasms that occurred 10 s after stunning were major disadvantages in the practical application of the method to veal calves, although they could be overcome by changing the pattern and site of application of the electrodes.

5.2.5. Sheep

In the principal sheep producing countries electricity is the most widely used means of pre-slaughter stunning and a variety of stunning devices is available. Three of the commonest patterns are (a) twin pronged electrodes, mounted on a pistol grip, that are applied to the back of the head, (b) a hand held unit with electrodes at either end, one of which is placed on the head and the other over the spine in the chest region, and (c) two handed scissor-type tongs, the electrodes of which are put on either side of the head. A restraining conveyor is used for stunning in the major meat works. This facilitates correct placing of the stunner and, with good design, enables sticking and shackling to be carried out with the minimum delay. The importance of this in relation to animal welfare has already been described in Section 5.2.1 and its relevance to meat quality is considered in Section 6.2. Smaller abattoirs commonly drive sheep into a pen for stunning. As mentioned in Section 3, this can make correct placing of the stunner difficult and, as the numbers of penned animals decrease, opportunities for flight increase and stress is exacerbated.

Conventional electric stunners operate over a range of about 70–110 V, 50 or 60 Hz. A stunning time of 7–10 s is commonly recommended, although application of the current is usually shorter in practice. Reference was made in Section 5.2.1 to the attempt by Leach *et al.*[55] to condition sheep to stunning with 90 V, 50 Hz, for periods between 3 and 15 s. All these sheep appeared to be satisfactorily stunned. However, failures to stun occur in practice. Two major factors are incorrect placement of the electrodes and the insulating effect of the wool.

Using between 90 and 130 V, 50 Hz, Kühne *et al.*[67] achieved satisfactory stunning of Karakul lambs with an initial current flow of 120–180 mA, increasing up to 350 mA over a 5 s period. They were unable to stun lambs with a current of less than 60 mA. Attempts have been made to improve stunning efficiency by using higher frequencies, a variety of wave forms and increased voltage. Most of this experimental work, already referred to in Section 5.2, had been carried out with pigs, however, in the course of Blackmore and Petersen's[53] observations on the stunning of calves, discussed in Section 5.2.4, they also studied sheep and lambs. They found it was necessary to induce cardiac dysfunction to ensure that the animals did not regain sensibility before slaughter in the existing system. This result could be achieved with a current of *c*. 0·8 A, 400 V applied for 3 s, either by the head-to-back or head-to-leg method. Using scissor type tongs applied to the head, Lambooy[68] estimated, on the basis of observations on the ECoG, that a current flow of 0·33 A would stun 90 % of sheep in a group weighing about 43 kg each. Although 600 V usually resulted in a longer epileptiform fit than 300 V, the shortest fits observed with both voltages were 23 and 22 s, respectively. He recommended, on animal welfare grounds, that a minimum of 0·5 A, 160 V should be used for sheep stunning and that sticking should take place within 16 s.

5.2.6. Pigs

Most slaughterhouses handling pigs use electrical pre-slaughter stunning, with restraining conveyors in the large works. In the course of their review of pre-slaughter stunning in the European Community, von Mickwitz and Leach[8] saw twelve different types of electrical stunning for this species. The diversity of equipment suggested that no consensus of opinion existed at that time on a satisfactory method of application. This conclusion was reinforced by the results of an investigation by Hoenderken[52] of pig stunning practice in twenty abattoirs, Section 2. Using as the criterion of a successful stun the onset of an electroplectic fit within one second of the start of the current flow, he determined statistically the amperage required to stun at least 90 % of the pigs. Variables included placing the electrodes of the scissor type tongs on one ear and the opposite eye or on both ears, wet or dry skin and well maintained or corroded electrodes. In six slaughterhouses using 70 V, 50 Hz, application on either the ear to eye or ear to ear position for 1 s failed to stun any pigs, while exposure for 5 s stunned between 13 % and 75 % of the 64 animals studied. The current flow varied between 0·1 and 0·4 A. Similar observations in nine slaughterhouses using 180 V, 50 Hz, showed that application by the ear to eye route was generally effective after either 1 or 5 s exposure. However, placing the electrodes in the ear to ear position was less reliable, a satisfactory stun being achieved in between 71 % and 100 % of the pigs. The current flow ranged between 0·3 and 1·3 A.

Stunners operating with 300–600 V ensured a satisfactory stun in most circumstances; current flow varied between 1·3 and 2·7 A. Using equipment supplying a constant current, Hoenderken found that, if 1·25 A flowed 0·5 s after the start of stunning, 98 % of the pigs would be stunned even with the tongs in the ear to ear position. To achieve this level of current flow, a minimum of 240 V was needed. Successful stunning caused a loss of sensibility for at least 32 s. These observations were of particular importance because they took account of the fact that the electrodes are not always well placed in practice and the findings are thus of particular relevance to the slaughterhouse situation.

A direct development from this work has been the introduction into commercial use of an automatic electrical stunner for pigs, described by Müller.[69] The electrodes are mounted in the path of animals travelling in a restraining conveyor.

5.2.7. Poultry

Electricity is in general use for pre-slaughter stunning of poultry. The two principal types of equipment are water bath and plate stunners. Most plants handling broilers use the former. On delivery at the slaughterhouse, the birds are suspended by their legs from shackles on a conveyor, which carries them to the stunner. The broilers' heads pass through an electrified saline solution for approximately 4–7 s and the current flows through their bodies to earth on the conveyor. The birds are moved on to an automatic throat cutter and then along a bleeding rail to the scalding tank. Water bath stunners are also used for large birds, but require adjustment to ensure that the bath is at the correct level for the size of the bird being handled. Problems can occur when this is not uniform.

In plate stunners the birds' heads come into contact with an electrified metal sheet. Some models have electrified spring wires projecting across the path of the birds. Plants handling spent hens and some processing turkeys use this pattern of stunner. Kuenzel and Ingling[70] found that a plate stunner was unsatisfactory for broilers because of inconsistent contact between the birds and the plate.

In view of the main current path in water bath stunning being through the body, it is not surprising that death may result from cardiac fibrillation. Van Mickwitz and Leach[8] found that 9 (28 %) of 32 broilers were dying when removed from the conveyor in a large commercial plant immediately after they left the stunner. Schütt-Abraham *et al.*,[63] working with broilers exposed for 4 s to 55–110 mA, 100 V, found 36 (43 %) of 84 examined after stunning had been killed by electrocution. Although bleed out was slowed from the dead birds, total blood loss differed insignificantly. Griffiths[71] compared bleed out from broilers killed by exposure to 240 V with that from birds stunned with 55, 80 or 105 V and then bled. The bleed out volume from all groups was similar, over 90 % of the total blood lost being collected within 2 min, when there was no significant difference between groups. The implication of these findings to poultry meat quality is discussed in Section 6.2.

5.3. Carbon Dioxide

A particular advantage of using carbon dioxide for immobilising animals before slaughter is that the problems of tissue residues, which are likely to occur if conventional anaesthetics are administered, do not arise, as it is a normal atmospheric constituent. Early investigations to determine the suitability of carbon dioxide for this purpose were made with a variety of laboratory animals by Bendersky,[72] but he concluded that the subject was more complex than it at first appeared.

Experimentation continued on the biochemical and physiological effects of inhaling varying concentrations of the gas. As a result of these studies, the responses to concentrations slightly higher than the normal carbon dioxide level in the atmosphere or to the relatively low concentrations used in surgical anaesthesia became well documented. However, scientific interest in the effects of the gas at concentrations of 60–80 %, as a means of pre-slaughter immobilisation, apparently remained in abeyance until the system was introduced into commercial practice for stunning pigs (Section 5.3.1).

Several workers have shown that sheep can be anaesthetised by high concentrations of carbon dioxide, but the economic disadvantage of the amount of gas removed on each animal's fleece has probably been a major factor in inhibiting application of this technique. In practice, therefore, it has been confined to pigs and, so far at an experimental level only, to poultry.

5.3.1. Pigs

As mentioned above, pre-slaughter immobilisation of pigs by carbon dioxide was first effected in a commercial abattoir. In the early 1950s the Hormel Company installed a plant capable of handling 1000 animals per hour in their works at Austin, Minnesota.[73] Pigs were carried on a conveyor down to the lowest level of the unit, where they were exposed to 60–65 % carbon dioxide in air for 45 s. On emerging from the plant, the pigs were transferred to a horizontal conveyor on which they were stuck and bled out.

After a visit to this abattoir, Wernberg[74] introduced the technique into Europe, first manufacturing an Oval Tunnel model, similar to the high capacity plant at Austin and current versions of which are capable of handling 120–600 pigs per hour.* A later development was the Compact model, operating on the same principle as a Ferris Wheel, with each pig entering an individual pen. The pens then move downwards into the carbon dioxide/air mixture and automatically eject the immobilised animals on their upward journey. These models have an hourly capacity of 90–300 pigs. A smaller model, the Dip-Lift, was introduced for abattoirs with a capacity up to 100 pigs per hour.

Research into the physiological effects of this technique was stimulated by its introduction into practice. Savić and Tadić[75] described the course of events in pigs immobilised by exposure to high concentrations of carbon dioxide. They found that a mixture of 60% carbon dioxide in air required prolonged exposure for the pigs to be immobilised, while increasing the carbon dioxide concentration over 80% did not reduce the speed of immobilisation significantly; they therefore selected a carbon dioxide concentration of 70% for their observations.

The animals showed no visible effect during the first 10–15 s of exposure to the gas mixture and then went through an excitement period, usually collapsing within 25–30 s of entering the chamber. The respiration rate fell sharply and the corneal reflex was abolished. This phase of light anaesthesia was followed by the animals lying still in deep anaesthesia. The corneal reflex returned 40–80 s after removal from the chamber and the pigs regained their feet within 2 or 3 min of this reflex being re-established. In a series of similar observations, using a Dip-Lift plant, Dodman,[76] working with carbon dioxide concentrations of 50% to 80% in air, confirmed Savić and Tadić's finding that the time taken for pigs to become recumbent significantly decreased as the carbon dioxide concentration increased. He noted that about half the pigs immobilised by exposure to 68–70% carbon dioxide in air for 45 s would be standing within 1 min of ejection from the plant.

Wellington *et al.*,[77] using a constant concentration of 75% carbon dioxide in oxygen and nitrogen, showed that, on exposure to the gas mixture administered by a mask, the respiration rate rose initially to 40–50 per min and then fell sharply. If exposed to the gas mixture for more than

* Butina ApS, 74C Gl. Kongevej, DK-1850 Copenhagen, Denmark.

5 min, respiratory failure was a contributory cause of death. These workers found that the rise in blood pressure which occurred on exposure to the gas mixture was not increased by raising the proportion of oxygen from 5 % to 25 %, which was in contrast to their results in sheep. Mullenax and Dougherty,[78] also working with gas mixtures administered by mask, found that arterial blood pressure of pigs rose during the first two minutes exposure to 68 % carbon dioxide in either oxygen or air; it was subsequently maintained in pigs receiving the carbon dioxide/oxygen mixture, but fell rapidly in those receiving carbon dioxide and air. Pigs given carbon dioxide with oxygen tended to collapse with less convulsive activity than those on the carbon dioxide/air mixture. The EEG was also monitored and the principal changes during inhalation of the gas mixture were an increase in amplitude, with some decrease in the high frequency component. The authors commented that the wave patterns did not permit a firm judgement concerning the anaesthetic state of the animals. No correlation was observed between convulsive seizures and the EEG trace.

The design of most of the plants in commercial use for carbon dioxide immobilisation does not permit routine observation of pigs during exposure to the gas mixture. Blomquist[79] reported on the reaction of pigs to 70 % carbon dioxide in air noted by an observer using breathing apparatus. During the first 15 s the pigs were apparently unaffected, they then went into a state of excitement and fell over. After 5 s their reaction to pain had apparently been lost or greatly reduced and 15 s after the onset of convulsions the corneal reflex disappeared. It returned about 80 s after removal of the animal from the gas mixture. Dodman[76] found the percentage of gas had no bearing on the incidence or duration of excitement. He noted that, while some pigs always had a relatively peaceful induction, others reacted violently at any concentration and he commented that the impression created was that the pigs were in distress.

Having trained groups of pigs to gain access to a chamber containing food, Cantieni[80] exposed them in this situation to 70 % carbon dioxide in air and assessed the acceptability of this method of immobilisation on the basis of the pigs' willingness to return to the food supply. He concluded from their subsequent hesitancy or refusal to re-enter the chamber that it did not correspond to the requirements of a good anaesthetic. Hoenderken *et al.*[81] studied the EEGs of pigs exposed to carbon dioxide concentrations in air ranging from 55–92 % (mean 75 $\pm$ 7·87 %). The mean durations of the behavioural stages were, 12 s initial phase, 26 s phase of excitement. The EEG trace first became abnormal after 25 s and iso-electric after 56 s. The authors concluded that the phase of violent excitation always started when the EEG was normal and the animal was therefore fully sensible. There was still brain activity at the start of insensibility during carbon dioxide administration, but the trace became iso-electric a few seconds later. On the basis of these results, The Netherlands Government in 1979 banned the use of carbon dioxide anaesthesia for pre-slaughter stunning. It is also proscribed in the Swedish Government's recently revised legislation (Fabiansson, personal communication; 1982).

Although protagonists of the use of carbon dioxide immobilisation for pre-slaughter stunning claim that sensibility is lost or much reduced within 5 s of the start of the phase of excitement[82] the fact that a normal EEG trace persists for several seconds while the pigs struggle with varying severity throws considerable doubt on the acceptability of this technique from the animal welfare point of view.

5.3.2. Poultry

Based on experimental observations of the effects on broilers of various carbon dioxide concentrations and exposure times, Kotula *et al.*[83] held the birds in 33–36 % carbon dioxide for 75 s before sticking, in a comparison of the rate and total volume of blood loss with that from similar birds that had not been immobilised before sticking. These workers subsequently developed an experimental plant designed to immobilise up to 4000 broilers per hour,[84] using the same gas concentration and exposure time. Birds were loaded individually onto a chute, which carried them to a conveyor running through the immobilising tunnel. On emerging, they were shackled before being stuck and bled out prior to scalding.

A US Department of Agriculture circular[85] described experimental work on the application of carbon dioxide to the immobilisation of turkeys. It was found that concentrations between 73 % and 77 % were satisfactory, but outside this range results were unreliable. Moreover, there was considerable variation in the exposure time necessary to immobilise birds of different varieties.

No further reports appear to have been published on the introduction of carbon dioxide immobilisation of poultry into commercial practice. A possible development, suggested by Kotula *et al.*,[84] would be modification of the tunnel to accept crates of birds directly from transport vehicles. Immobilisation within the crates should facilitate subsequent removal and shackling, operations that are recognised to be a major source of damage occurring in existing plants.

There is a lack of experimental evidence on the physiological reactions of poultry to high levels of carbon dioxide. On the basis of experience gained from the application of this technique of immobilisation to mammals, thorough investigation of its effects on the avian central nervous system is required before advocating large scale use of carbon dioxide immobilisation in poultry slaughterhouses.

6. DEFECTS CONNECTED WITH PRE-SLAUGHTER STUNNING

6.1. Fractures and Bruising

In discussing skeletal fractures resulting from electrical stunning of 100 kg pigs, van der Wal[86] pointed out that they not only caused economic loss through the need to trim tissues stained by bleeding, but also put cutting room staff at risk of injury by bone splinters. The scapula was most often affected, fractures occurring in the spine, neck or glenoid cavity. From an analysis of the numbers observed in several slaughterhouses, van der Wal found 1·5 % of pigs stunned using 70 V had shoulder fractures, whereas they were present in 2·2 % of pigs stunned by 180 V and an incidence of over 10 % had been recorded in extreme cases. However, if the pigs were stunned in a restraining conveyor, no shoulder fractures occurred. He concluded that fractures were attributable to the initial violent extension of the limbs when the stunning current was applied. The feet of pigs in a restraining conveyor were not in contact with the floor, so that their limbs were not exposed to the counter pressure experienced by free standing animals. Fractures of the spine, usually affecting the fifth or sixth thoracic vertebra, and of the neck of the femur occurred infrequently. However, since they affected the *M. longissimus dorsi* or the ham, the loss of value could be important in individual carcasses.

There appear to be no reports of similar fractures occurring in sheep or lambs, although large numbers are stunned electrically with equipment similar to that used for pigs.

Bruising of stock, particularly cattle, during transport is a well recognised source of loss. It has been shown that bruising can occur until blood pressure has fallen almost to zero[87] and Meischke and Horder[88] studied the degree of bruising occurring in cattle on ejection from stunning boxes. Animals fell about 60 cm and their manner of fall was noted. The severity and site of bruises were recorded during dressing and the results from each side were compared. The degree of bruising was affected by the way in which animals fell, but those falling completely onto one side showed a highly significant difference between the amount of bruising on their two sides. This was also the case, though to a lesser degree, with animals falling less cleanly. The only exceptions were animals that slid out of the box, in which there was no significant difference between the two sides.

6.2. Blood Splash

Intramuscular haemorrhages, which may be of limited distribution or widespread and can vary in size between pinheads (petechiae) to more than a centimetre in diameter (ecchymoses), are a well recognised defect seen sporadically in the carcasses of all species of meat animals after pre-slaughter stunning.

Their occurrence has been linked, in particular, with the introduction of electrical stunning. In 1955 an official committee on slaughtering methods[89] was set up in Denmark to investigate the effects of electrical stunning on pork quality and its possible association with haemorrhages in the muscles. The committee reported that the problem of blood splash could be avoided by reducing the interval between the end of stunning and sticking to about 5 s.

Arterial blood pressure rises steeply during stunning and a further increase occurs after current flow ceases in electrical stunning. However, the supposition that capillaries are ruptured by an onrush of blood from the arterioles when vasodilation occurs at the end of electrical stunning[90] is scarcely tenable when the extent and capacity of the capillary bed is considered. Moreover, Mandrup[91] demonstrated that the secondary increase in blood pressure immediately after electrical stunning was not responsible for blood splash by exsanguinating groups of pigs during and after electrical stunning. The levels of blood splash in both groups were similar.

A major development in the understanding of blood splash was the demonstration by Leet *et al.*[92] that about 5 % of the muscle fibres in blood splash areas were distorted, with zones of extreme shortening irregularly interspersed with less severely affected zones. This phenomenon could be reduced by neuromuscular blocking agents. These workers proposed that supercontraction of some fibres puts a heavy strain on adjacent capillaries, sometimes causing their walls to rupture, enabling haemorrhage into the tissues to occur.

The importance of prompt sticking of lambs was demonstrated by Kirton *et al.*[93] who found a lower level of blood splash in lambs exsanguinated 0·5–2 s after electrical stunning compared with lambs exsanguinated after 4–8 s. They suggested that application of electrical current to the heads of lambs resulted in damage to small blood vessels before the secondary rise in blood pressure, which then exacerbated the leakage of blood, giving rise to discrete haemorrhages. Subsequently, Kirton *et al.*[94] found that the incidence of carcass blood splash was much lower in lambs stunned by the head-to-back method than on the head only. By measuring blood loss from lambs stunned electrically on the head or by the head-to-back method, compared with that lost from lambs bled out without stunning, these authors found that similar quantities of blood were collected from the head stunned and unstunned lambs over 120 s. Although head-to-back stunned lambs only yielded 46 % of the weight of blood recorded from head stunned lambs, little or any of the blood not collected remained in the carcass after dressing. A similar result was reported by Chrystall *et al.*[95] who found that, although there were differences in the amount of blood lost on sticking, there were no differences in the residual blood content of the *M. longissimus dorsi*, the microbiological status of the meat or its tenderness, in a study of various stunning procedures in sheep.

In a comparison of blood splash levels in lambs after different methods of pre-slaughter stunning, Kirton *et al.*[96] found no commercially significant levels in the carcasses. Detectable splashing was found in 25 % of the carcasses after electrical stunning and only 1 % after concussion stunning, with captive bolt stunning in an intermediate position.

Hatton and Ratcliff[97] compared the incidence of blood splash in the shoulder joints of pigs stunned with 95–100 V, 550–750 mA, at 50 Hz or 1300 Hz. Exposure to 50 Hz current for 6–9 s resulted in 60–80 % haemorrhages; reducing the time to 3 s lowered the level to 40 %. An application of 1300 Hz for 9 s brought down the incidence to 20 % and it was reduced to $<10\%$ by exposure for 3 s.

The occurrence of an unusually high level of blood splash in a group of lambs led Restall[98] to examine the flock of origin by the one-stage prothrombin test. Samples from control lambs on a commercial slaughter line yielded 35·4 % with extended prothrombin times, while they occurred in all samples from the flock under test. The pasture grazed by the affected flock contained coumarin producing plants. As some of this group of drugs prolong one-stage prothrombin times and induce capillary fragility, it was postulated that their ingestion might result in increased levels of blood splash.

Although blood splash is generally associated with haemorrhage into the musculature, Thornton *et al.*[99] described the occurrence of pinpoint haemorrhages in the carcass fat, a condition termed 'speckling'. They stated that it was particularly prevalent in young lambs and most noticeable in lean carcasses. It appeared to occur irrespective of the type of electrical stunning used. Blackmore and Petersen[53] found the prevalence of speckling was increased when the electrodes only contacted the hind legs in head-to-leg stunning. They suggested that failure to contact the forelegs resulted in less effective cardiac inhibition and an increase in current flowing between the head and hind legs, through the fatty tissue of the loin and hind leg where speckling is most often detected.

6.3. Other Meat Quality Defects

Discussing the effect of stunning methods and slaughter on meat quality, Marple[100] stated that the type of stunning had very little influence on the ultimate quality of lamb or beef carcasses and he confined his subsequent comments to its effect on pork. It is undeniable that research has tended to focus on meat quality defects in pigs, probably because of their more labile biochemical response to the stresses inseparable from transport and slaughter. There is very little evidence to suggest that stunning and slaughter methods *per se* affect the organoleptic characteristics of meat from ruminants.

In a study of meat quality in two groups of pigs subjected to different types of electrical stunning, van der Wal *et al.*[101] noted the pH and internal temperature of the *M. semimembranosus* and the *M. longissimus dorsi*, as well as the degree of rigor mortis in the former, 45 min after slaughter. They also carried out a sensory evaluation 24 h post-slaughter. The stunning methods compared were in use in two slaughterhouses that drew their pigs from different areas. In one plant the pigs were stunned with 300 V, applied by hand held tongs, and in the other they were stunned, in an automated unit, with 680 V. No clear evidence was found that the two methods affected meat quality differently. The authors stressed that the number of factors recognised to affect meat quality, e.g. genetic predisposition, nutrition, transport and lairing, made it extremely difficult to isolate factors in stunning technique that were responsible for defects in meat quality.

Nevertheless, Leest *et al.*[102] showed that pigs immobilised with carbon dioxide had lower pH_1 values than pigs stunned with 70 V or 300 V; while Sybesma and Groen[103] found that pigs passed through a carbon dioxide oval tunnel, without the gas being present, and then stunned with 70 V had a more rapid post-mortem pH fall than pigs not exposed to the tunnel, but also stunned with 70 V.

Damage to poultry carcasses during processing is a cause of appreciable loss, but is seldom attributable to the stunning and bleeding operations. There is a tendency to use low voltages in water bath stunners, in the belief that heart action is less likely to be impaired, so that effective bleeding out will result. Reference has already been made (Section 5.2.7) to the fact that birds in commercial plants not infrequently die from syncope after water bath stunning without their carcasses being detectably impaired and, consequently, without the knowledge of the management. Heath *et al.*[104] found that birds stunned with *c.* 120 V yielded carcasses indistinguishable by the factory staff from birds stunned with <90 V, which was routinely applied in the plant. In a study of the effect of raising the voltage in water bath stunners, Veerkamp and de Vries[105] showed there was a decrease in bleed out with increase in voltage; 90 s after throat cutting the proportion of total blood lost by birds stunned with 75 V was 35·08 %, against 28·07 % in birds stunned with 200 V. They also found colour differences were discernible in the wing tips and pygostyles. According to Heath *et al.*,[104] redness in these areas is a result of the large flight feathers being torn from their attachments and is not related to bad bleeding. As mentioned previously (Section 5.2.2), Griffiths[71] found that stunning in a water bath with up to 240 V did not affect carcass quality.

Birds that are bright red after scalding and plucking are known as 'redskins' and the condition is popularly attributed to defective bleeding. Examination of a number of these carcasses by Heath *et al.*[104] showed that none had been exsanguinated. However, birds that had been intentionally killed by high voltage stunning and then processed without being bled out gave white carcasses. When the skin of redskin carcasses was reflected, the subcutaneous tissues were white and the colour, presumably due to erythrocytes or haemoglobin, was confined to the dermis. The redskin birds almost always had swollen heads, caused by blood stained fluid in the subcutaneous tissues, while this condition was never seen in birds killed but not bled before scalding. It was concluded that the redskin condition probably resulted from birds entering the scalder while still alive.

7. CONCLUSIONS

7.1. Pre-slaughter Stunning Methods

Research during the last decade, in particular the application of electroencephalography to monitoring pre-slaughter stunning practices, has provided a firmer basis for assessing their effectiveness.[26] Captive bolt stunning is an acceptable method for cattle[28] and sheep,[38] provided cartridges of adequate strength are used[36] and the pistol is correctly placed. Although 200 kg calves can be shot satisfactorily either in the frontal region or at the occiput,[29] there is insufficient information at present on the reliability of the latter site for shooting adult cattle. Sheep should be shot in the occipital region. On anatomical grounds, the captive bolt pistol is not recommended for stunning any but light weight pigs.

The available evidence suggests that concussion stunning is not altogether satisfactory[40] and there is a dearth of scientific data on its use in adult cattle and pigs.

Electrical stunning has received most attention from research workers, which is perhaps an indication that it is the technique raising the most queries, particularly in relation to the occurrence of muscle paralysis without loss of sensibility.[54] The demonstration that it is necessary to use 1·1 A to effect a reliable stun in the majority of 100 kg pigs[52] under slaughterhouse conditions has stimulated a reappraisal of electrical stunning techniques. An adequate current can provoke an electroplectic fit in milliseconds, which raises doubts about the acceptability of methods requiring current to flow for several seconds to achieve an effective stun. Although high frequencies have advantages in reducing the prevalence of blood splash in pigs,[97] some doubt exists about their painlessness.[24] Research with pigs,[60] sheep[62] and poultry[63] has shown that it is possible to put the heart into fibrillation during the act of stunning without impairing carcass quality. The cessation of oxygen supply to the brain ensures that the animal will not recover sensibility at the conclusion of the electrical stun. This is an important development in stunning technology and it should be stressed that it is based on the response of physiologically normal animals. However, in situations where electrical stunning is accepted in conjunction with religious slaughter methods, on the understanding that the animal would recover completely if not bled out, application of the electrodes must be confined to the head to avoid the risk of initiating cardiac fibrillation.

Controversy surrounds the acceptability of carbon dioxide immobilisation as a pre-slaughter stunning method for pigs.[81,82,106] Recent work suggests that the animals are subjected to a fairly severe stress period before loss of sensibility occurs.[76,107] However, the high initial cost of the plant and expense of maintenance, compared with other stunning methods, as well as the lack of compelling advantages, suggest that economic considerations militate against it being widely adopted.

Maintenance of stunning equipment is important. Captive bolts can become distorted or corroded, with consequent loss of power. Electrical stunning equipment should always incorporate ammeters and voltmeters, so that current flow can be frequently monitored. A system of automatic alarms is required to signal variations in carbon dioxide concentrations beyond set limits. There is much to commend the requirement operating in some countries[8] that any new pattern of stunning equipment should be subject to official testing before approval for general use.

7.2. Stunning Facilities

In countries where beef cattle are stall fed and so accustomed to handling, they can be stunned in the slaughterhall with the minimum of restraint and, as the slaughterman is working in comfortable conditions, he can operate the stunning instrument with ease and accuracy. Where cattle are raised extensively it is necessary to confine them in a stunning box or restraining conveyor. The latter system presents few problems, as the slaughterman can be positioned to stun correctly. However, the design of many stunning boxes takes little account either of animal behaviour or working conditions. It is consequently difficult for the slaughterman to operate correctly and faulty cattle stunning is largely attributable to poor box design.[8] Another factor is the requirement in some national legislations that an animal must not be killed in the sight of another. Careful observation suggests that there is less distress if animals are stunned in the presence of others than if subjected to the unusual experience of isolation. In this connection, the following extract from the report of a United Kingdom government committee[108] is noteworthy:

> 'although it might be argued that for one animal to watch another being stunned and slaughtered must cause alarm and stress, our observations have convinced us that this is not the case'.

Clearly the amendment of such anthropocentric legislation would be difficult, but it would encourage a fresh approach to stunning box design, with consequent benefit to animal welfare and working efficiency.

The problems associated with stunning pigs and sheep in a pen were alluded to in Section 3. Some slaughterhouses use a gate to reduce the floor area as the number of animals decreases, others have a hydraulically operated mobile wall mounted on one side of the pen that serves the same purpose. There is a need in slaughterhouses with a comparatively low throughput for a simple design of race that would enable small stock to be immobilised and stunned accurately and quickly.

The possibility that electrically stunned animals may recover sensibility before exsanguination is a well-recognised source of concern.[31] Apart from the animal welfare aspect, it is also important in relation to the occurrence of blood splash.[89] This risk can be avoided by plant designs that enable the animals to be stuck in the horizontal position, before shackling and hoisting. Where this not possible, attention should be given to the speed of the elevator. As already indicated [31,53,63] this problem can be avoided by using stunning techniques that also cause cardiac fibrillation.

7.3. Legislation

Commenting on the United Kingdom legislation, Everton [13] emphasised its lack of precision and the need to take account of advances in scientific knowledge. These criticisms could be directed at the corresponding laws in most developed countries. It is noteworthy that The Netherlands Government has recently revised its regulations on electrical stunning (Lambooy, 1983; personal communication), requiring:

(a) that an alternating current of 50 Hz must flow straight through the brain,
(b) the potential between the electrodes must be at least 240 V during stunning and limits are set for the maximum voltage according to the type of equipment. The no-current potential must not exceed 24 V,
(c) animals must be bled within 30 s of stunning.

The United Kingdom's original official circular [10] on electrical stunning recommended applying a minimum of 75 V for at least 7 s. Although successful stunning can be achieved with this voltage by careful electrode placement, it gives little margin for error.[52] In slaughterhouse practice, the current is seldom applied for more than 2 or 3 s and doubts about the acceptability of using currents that require longer exposure times have been expressed under Section 7.1.

Standards of pre-slaughter stunning would undoubtedly be raised by closer, informed supervision and a clearer definition of responsibility for this role would perhaps be the greatest contribution the legislation could make to the improvement of stunning practice.

The hazards to which slaughtermen are exposed by poor stunning box design have already been touched on. A more recent concern in respect of worker safety is the introduction of high voltage stunning. A paper by Kamphuis [109] discussed the hazards involved and the available means of protection.

7.4. Training

With the exception of automated high voltage stunning and carbon dioxide immobilisation, successful application of the techniques in use at present is dependent on the skill of the slaughterman. It is clearly important that training should be given both in the principles and practices involved. This can be achieved by including training requirements in the licensing of slaughtermen. However, the United Kingdom system of stipulating a period under the instruction of an experienced slaughterman tends to ensure that faulty practices pass from one generation to the next. To be effective, it is necessary for at least part of the training to be given in technical schools.

8. REFERENCES

1. Robinson, A. J. (1973). *Calf stunning*, Meat Industry Research Institute of New Zealand publication 292. Hamilton, New Zealand.
2. Thornton, H. (1973). *Aspects of meat inspection*, Ballière Tindall, London, p. 24.
3. Blackmore, D. K. and Newhook, J. C. (1983). In: *Stunning animals for slaughter* (Eikelenboom, G. (Ed.)), Martinus Nijhoff, The Hague, p. 13.
4. *Oxford English Dictionary* (1970). Vol. VII. Clarendon Press, Oxford.
5. Rasenack (1979). *Fleischwirts*, **50**, 612.
6. Fahrbach, R. (1948). Doctoral dissertation, Tierärztliche Hochschule, Hannover.
7. *Slaughterhouse Act, 1974*. Her Majesty's Stationery Office, London.
8. Mickwitz, G. von and Leach, T. M. (1977). *Review of pre-slaughter stunning in the European Community*. Information on Agriculture No. 30. Office of Official Publications of the European Economic Communities, Luxembourg.
9. *Handboek Vleesverwekende bedrijven*. Deel I, Varkensslachterijen. Den Haag, Min. v. Landbouw enz., zj.
10. *Slaughter of Animals Acts—Electric Stunning (1958)*. Circular FSH3/58. Her Majesty's Stationery Office, London.
11. *Slaugherhouses Act, 1974: Electrical Stunning*. Circular FSH5/78. Her Majesty's Stationery Office, London.
12. *Midas Bulletin No. 4 (1978). Stunning and bleeding cattle, sheep and pigs*. Meat and Livestock Commission, Bletchley, United Kingdom.
13. Everton, A. R. (1982). *Vet. Rec.*, **110**, 469.
14. Wal, P. G. van der (1978). *Meat Sci.*, **2**, 19.
15. Lister, D., Gregory, N. G. and Warriss, P. D. (1981). In: *Developments in Meat Science—2* (Lawrie, R. A. (Ed.)), Applied Science Publishers, London, p. 61.
16. Grandin, T. (1980). *Int. J. Stud. Anim. Prob.*, **1**, 121.
17. Grandin, T. (1980). *Int. J. Stud. Anim. Prob.*, **1**, 178.
18. Blackmore, D. K. (1983). In: *Stunning animals for slaughter* (Eikelenboom, G. (Ed.)), Martinus Nijhoff, The Hague, p. 167.
19. Kitchell, R. L. (1971). In: *The assessment of pain in man and animals* (Keele, C. A. and Smith, R. (Eds)), Universities' Federation of Animal Welfare, London, p. 244.

20. PRICE, H. L. (1971). In: *Textbook of veterinary anaesthesia* (Soma, L. R. (Ed.)), Williams and Wilkins, Baltimore, p. 108.
21. ROOS, J. and KOOPMANS, S. (1934). *Vet. J.*, **90**, 232.
22. ROOS, J. and KOOPMANS, S. (1936). *Vet. J.*, **92**, 127.
23. HALL, L. W. (1971). *Wright's veterinary anaesthesia and analgesia*, 7th ed., Baillière Tindall, London, p. 199.
24. CROFT, P. G. (1952). *Vet. Rec.*, **64**, 225.
25. CROFT, P. G. (1952). *J. Ment. Sci.*, **98**, 427.
26. LOPES DA SILVA, F. H. (1983). In: *Stunning of animals for slaughter* (Eikelenboom, G. (Ed.)), Martinus Nijhoff, The Hague, p. 3.
27. RUCKENBUSCH, Y. (1965). *Proc. R. Soc. Med.*, **58**, 551.
28. FRICKER, C. and RIEK, W. (1981). *Fleischwirts.*, **61**, 124.
29. LAMBOOY, E. and SPANJAARD, W. (1981). *Vet. Rec.*, **109**, 359.
30. NEWHOOK, J. C. and BLACKMORE, D. K. (1982). *Meat Sci.*, **6**, 221.
31. GREGORY, N. G., WOTTON, S. B. and LISTER, D. (1983). In: *Stunning animals for slaughter* (Eikelenboom, G. (Ed.)), Martinus Nijhoff, The Hague, p. 123.
32. OLIVER, D. F. (1979). *Vet. Rec.*, **105**, 224.
33. BUTLER, E. G., PUCKETT, W. A., NEWTON HARVEY, E. and MCMILLEN, J. H. (1945). *J. Neurosurg.*, **2**, 358.
34. HOPKINSON, D. A. W. and MARSHALL, T. K. (1967). *Brit. J. Surg.*, **54**, 344.
35. LAMBOOY, E. (1981). *Fleischwirts.*, **61**, 1882.
36. LAMBOOY, E., LOGTESTIJN, J. G. VAN and SYBESMA, W. (1983). In: *Stunning animals for slaughter* (Eikelenboom, G. (Ed.)), Martinus Nijhoff, The Hague, p. 138.
37. GROSS, R. (1976). Doctoral dissertation, Tierärztliche Hochschule, Hannover.
38. FREESEMANN, L. (1975). Doctoral dissertation, Tierärztliche Hochschule, Hannover.
39. LAMBOOY, E. (1982). *Meat Sci.*, **7**, 51.
40. SCHÜTT-ABRAHAM, I., WORMUTH, H.-J., FESSEL, J. and KNAPP, J. (1983). In: *Stunning animals for slaughter* (Eikelenboom, G. (Ed.)), Martinus Nijhoff, The Hague, p. 154.
41. PARKER, T. (1929). *Vet. J.*, **85**, 197.
42. HOLBOURN, A. H. S. (1945). *Brit. Med. Bull.*, **3**, 147.
43. LAMBOOY, E. and EIKELENBOOM, G. (1983). In: *Stunning animals for slaughter* (Eikelenboom, G. (Ed.)), Martinus Nijhoff, The Hague, p. 134.
44. WALKER, A. E., KOLLROS, J. J. and CASE, T. J. (1955). *J. Neurosurg.*, **1**, 103.
45. NILSSON, B. and PONTÉN, U. (1977). *J. Neurosurg.*, **47**, 252.
46. BLACKMORE, D. K. (1979). *Vet. Rec.*, **105**, 372.
47. LAMBOOY, E., SPANJAARD, W. and EIKELENBOOM, G. (1981). *Fleischwirts.*, **61**, 128.
48. WARRINGTON, R. (1974). *Vet. Bull.*, **44**, 617.
49. SZÜCS, A., BOHN, S. and SZABADOS, A. (1963). *Proc. 9th Conf. Euro. Meat Res. Wkrs*, Budapest.
50. KOLEDIN, I. D. (1963). *Proc. 9th Conf. Euro. Meat Res. Wkrs*, Budapest.
51. BORZUTA, K. (1971). *Roczn. Instytutu Prezmyslu Miesnego*, **8**, 27.
52. HOENDERKEN, R. (1979). Doctoral dissertation, Rijksuniversiteit te Utrecht.
53. BLACKMORE, D. K. and PETERSEN, G. V. (1981). *New Zealand Vet. J.*, **29**, 99.
54. HUME, C. W. (1935). *Nature*, **135**, 107.

55. LEACH, T. M., WARRINGTON, R. and WOTTON, S. B. (1980). *Meat Sci.*, **4**, 203.
56. MÜLLER, A. and ENGELI, P. (1968). *Fleischwirts.*, **48**, 1346.
57. GREGORY, N. G. and WOTTON, S. B. (1983). *Res. Vet. Sci.*, **34**, 315.
58. WARRISS, P. D. (1977). *J. Sci. Fd Agric.*, **28**, 457.
59. WARRISS, P. D. and LEACH, T. M. (1978). *J. Sci. Fd Agric.*, **29**, 608.
60. WARRISS, P. D. and WOTTON, S. B. (1981). *Res. Vet. Sci.*, **31**, 82.
61. SWATLAND, H. J. (1983). *Canad. Inst. Fd Sci. Tech. J.*, **16**, 35.
62. KIRTON, A. H., FRAZERHURST, L. F., WOODS, E. G. and CHRYSTALL, B. B. (1981). *Meat Sci.*, **5**, 347.
63. SCHÜTT-ABRAHAM, I., WORMUTH, H.-J. and FESSEL, J. (1983). In: *Stunning animals for slaughter* (Eikelenboom, G. (Ed.)), Martinus Nijhoff, The Hague, p. 187.
64. ANON. (1959). *Fleischwirts.*, **11**, 370.
65. NEWHOOK, J. C. and BLACKMORE, D. K. (1982). *Meat Sci.*, **6**, 295.
66. LAMBOOY, E. and SPANJAARD, W. (1982). *Meat Sci.*, **6**, 15.
67. KÜHNE, K. J., JENKINS, W. L. and KRUGER, J. M. (1979). *J. South Afr. Vet. Assn*, **50**, 15.
68. LAMBOOY, E. (1982). *Meat Sci.*, **6**, 123.
69. MÜLLER, J. (1981). *Fleischwirts.*, **61**, 1518.
70. KUENZEL, W. J. and INGLING, A. L. (1977). *Poultry Sci.*, **56**, 2087.
71. GRIFFITHS, G. L. (1983). *Proc. Tech. Seminar, Far East and South Pacific Fed., Wld Poultry Sci. Assn*, Adelaide, p. 10.
72. BENDERSKY, M. J. (1904). *C.R. Soc. Biol.*, **57**, 458.
73. SWEM, E. R. (1952). *Nat. Provis.*, **127**, 8.
74. WERNBERG, N. E. (1955). *Konserves*, **13**, 32.
75. SAVIĆ, I. and TADIĆ, R. (1963). *Proc. 9th Conf. Europ. Meat Res. Wkrs*, Budapest.
76. DODMAN, N. H. (1977). *Br. vet. J.*, **133**, 71.
77. WELLINGTON, G. H., DAVIS, W. F., STOUFFER, J. R., WENTWORTH, R. A. and BARR, G. R. (1964). *Proc. 16th Res. Conf. Am. Meat Inst. Fndn*, University of Chicago, p. 107.
78. MULLENAX, C. H. and DOUGHERTY, R. W. (1964). *Am. J. Vet. Res.*, **25**, 424.
79. BLOMQUIST, S. M. (1957). *Food Manufacture*, **32**, 230.
80. CANTIENI, J. (1977). *Schweitz. Arch. Tierheilk*, **119**, 355.
81. HOENDERKEN, R., LOGTESTIJN, J. G. VAN, SYBESMA, W. and SPANJAARD, W. J. M. (1979). *Fleischwirts.*, **59**, 1572.
82. LOMHOLT, N. (1983). *Dtsch. tierärztl. Wschr.*, **90**, 66.
83. KOTULA, A. W., DREWNIAK, E. E. and DAVIS, L. L. (1957). *Poultry Sci.*, **36**, 585.
84. KOTULA, A. W., DREWNIAK, E. E. and DAVIS, L. L. (1961). *Poultry Sci.*, **40**, 213.
85. DREWNIAK, E. E., BAUSH, E. R. and DAVIS, L. L. (1955). Circular 958, U.S. Department of Agriculture, Washington, D.C.
86. WAL, P. G. VAN DER (1976). *Proc. 22nd Euro. Meet. Meat. Res. Wkrs*, Malmö.
87. HAMDY, M. F., KUNKLE, L. E. and DEATHERAGE, F. E. (1957). *J. Anim. Sci.*, **16**, 490.
88. MEISCHKE, H. R. C. and HORDER, J. C. (1976). *Fd Technol. Aust.*, **28**, 369.

89. COMMITTEE ON SLAUGHTERING METHODS (1959). Danish Academy of Technical Sciences, Copenhagen.
90. THORNTON, H. and GRACEY, J. F. (1974). *Textbook of Meat Hygiene*, 6th Ed., Baillière Tindall, London.
91. MANDRUP, M. (1964). *Proc. 10th Meet. Euro. Meat Res. Wkrs*, Roskilde.
92. LEET, N. G., DEVINE, C. E. and GAVEY, A. B. (1977). *Meat. Sci.*, **1**, 229.
93. KIRTON, A. H., BISHOP, W. H., MULLORD, M. M. and FRAZERHURST, L. E. (1978). *Meat Sci.*, **2**, 199.
94. KIRTON, A. H., FRAZERHURST, L. F., WOODS, E. G. and CHRYSTALL, B. B. (1981). *Meat Sci.*, **5**, 347.
95. CHRYSTALL, B. B., DEVINE, D. E. and NEWTON, K. G. (1981). *Meat Sci.*, **5**, 339.
96. KIRTON, A. H., FRAZERHURST, L. F., BISHOP, W. H. and WINN, G. W. (1981). *Meat Sci.*, **5**, 407.
97. HATTON, M. and RATCLIFFE, P. W. (1973). *Proc. 19th Meet. Europ. Meat Res. Wkrs*, Paris.
98. RESTALL, D. J. (1981). *Meat Sci.*, **5**, 125.
99. THORNTON, R. N., BLACKMORE, D. K., JOLLY, R. D., HARRIS, R. E. and MARSDEN, N. A. (1979). *N.Z. Vet. J.*, **27**, 181.
100. MARPLE, D. N. (1977). *Proc. Meat Ind. Res. Conf.*, Amer. Meat Ind. Fndn, Arlington.
101. WAL, P. G. VAN DER, EIKELENBOOM, G. and LAMBOOY, E. (1983). In: *Stunning animals for slaughter* (Eikelenboom, G. (Ed.)), Martinus Nijhoff, The Hague, p. 82.
102. LEEST, J. A., ROON, P. S. VAN and BROUWER, H. A. (1970). *Proc. 16th Meet. Euro. Meat Res. Wkrs*, Varna.
103. SYBESMA, W. and GROEN, W. (1970). *Proc. 16th Meet. Euro. Meat. Res. Wkrs*, Varna.
104. HEATH, G. B. S., WATT, D. J., WAITE, P. R. and MEAKINS, P. A. (1983). *Br. Vet. J.*, **139**, 285.
105. VEERCAMP, C. H. and VRIES, A. W. DE (1983). In: *Stunning of animals for slaughter* (Eikelenboom, G. (Ed.)), Martinus Nijhoff, The Hague, p. 197.
106. LAURSEN, A. MOSFELDT (1983). In: *Stunning of animals for slaughter* (Eikelenboom, G. (Ed.)), Martinus Nijhoff, The Hague, p. 64.
107. LEACH, T. M. and WARRINGTON, R. (1976). *Med. Biol. Eng.*, **14**, 79.
108. COMMITTEE ON THE EXPORT OF ANIMALS FOR SLAUGHTER (1974). Cmnd. 5566. Her Majesty's Stationery Office, London, p. 20.
109. KAMPHUIS, L. (1983). In: *Stunning of animals for slaughter* (Eikelenboom, G. (Ed.)), Martinus Nijhoff, The Hague, p. 35.

Chapter 4

PACKAGING FRESH MEAT

A. A. TAYLOR

AFRC Food Research Institute,
Langford, Bristol, UK

1. INTRODUCTION

The development of meat packaging during the last 20 years has brought about major changes in the pattern of distribution and marketing. At wholesale level, the traditional movement of beef carcasses or part-carcasses between abattoir and retail outlet has largely been replaced by the distribution from cutting and packing plants of boxed, boneless joints, vacuum packed to give long life without weight loss. Moreover, these partly prepared joints are convenient to use and economical in terms of labour and space. At retail level, supermarket and self-service selling of fresh and frozen meat is possible only with correctly used packaging. In-store packing of meat cuts is being supplemented, and may eventually be replaced completely, by the distribution of pre-packed retail cuts from centrally located factories. As this has developed, the limitations of conventional methods of packing have become apparent. New techniques are being devised to extend the display life of fresh meat. In this review, the relative merits of different approaches are considered, not only for chilled meat, but also for frozen meat which may become an economically sound alternative means of retailing.

2. BEHAVIOUR OF MEAT

The effects of different packaging systems are best studied by first considering the behavioural properties of meat which are directly influenced.

2.1. Water Loss

In most cases, uncovered meat loses weight by evaporation of moisture which dries and darkens the surface. During frozen storage water is lost from the meat surface by sublimation to colder surfaces in the vicinity. These may be refrigeration coils, walls or inside surfaces of containers, and again weight loss is accompanied by surface discoloration. Water is also lost as the principal component of the exudate which inevitably comes from the cut surfaces of chilled meat.

2.2. Tissue Respiration

Meat is a biological material containing respiratory enzyme systems which continue to function after death. They consume oxygen and produce carbon dioxide, at a fast rate for 1–2 days, and thereafter at a slower rate. The level of respiration varies with muscles and animals and also between species.

Respiration is confined to the surface layer into which oxygen diffuses. The depth of penetration depends on a balance between oxygen concentration at the surface, driving it inwards, and tissue respiration which consumes the oxygen as it becomes available. Consumption is less and solubility greater at lower temperatures and therefore depth of penetration is greater. Aerobic bacteria on the surface of the meat may also consume oxygen and produce carbon dioxide. While the bacterial load is low, the gases involved may be negligible in quantity, but during storage, spoilage may increase to a level where microbial respiration is comparable with that of muscle tissue.

2.3. Microbiology

After dressing, the surface of a beef carcass may carry between 10^2 and 10^4 bacteria/cm^2, and after butchery, joints and pieces of meat for packing are likely to carry considerably higher numbers. They may comprise a large range of different bacteria some of which can grow between 0 and 5 °C. Among these are *Pseudomonas* spp. which can multiply in air and become the most dominant bacteria on chilled meat. Their proliferation eventually causes slime formation and the putrid odours commonly associated with spoiled meat.

Pseudomonas spp. require oxygen for growth and, in air, will spoil meat in 10 days at 0 °C or 5 days at 5 °C. Other micro-organisms including lactic acid bacteria also grow on beef stored in air, but they are rapidly outgrown by the pseudomonads and contribute little to spoilage of unprotected meat.

A detailed review of microbiological and chemical changes in chilled stored meats has been published.[1]

2.4. Colour

Lean meat colour derives from the muscle pigment, myoglobin, and its reactions with available oxygen. The depth (X) of oxygen penetration into a piece of meat is governed by the partial pressure (C_0) of the gas at the surface, the rate of oxygen consumption (A) by the tissue and the diffusion constant (D), according to the equation of Brooks:[2]

$$X = \sqrt{\frac{2C_0D}{A}}$$

In meat which has been exposed to air for several hours the penetration depth may be 6–7 mm. In the presence of oxygen, myoglobin is oxygenated to oxymyoglobin, or oxidised to metmyoglobin. The relative amount of these two pigment forms depends on the partial pressure of oxygen. Oxymyoglobin, the bright-red ferrous form of the pigment, is favoured by high oxygen concentration, while low concentrations favour oxidation to brown ferric metmyoglobin. The optimum condition for oxidation is approximately 1 mm oxygen partial pressure. When a freshly cut piece of meat is exposed to air, both reactions take place; at the surface, where oxygen is freely available, oxymyoglobin forms and extends inwards to near the limit of oxygen penetration where metmyoglobin predominates. Beyond this, where oxygen has not penetrated, the pigment remains as purple, reduced myoglobin.

Oxygenation of myoglobin is rapid; the surface of beef in air will appear red within half an hour at 5 °C. On the other hand, oxidation to metmyoglobin is slow, and appears first as a thin brown layer at the limit of oxygen penetration. Within 1 or 2 days this layer thickens and becomes apparent, first by darkening the translucent surface tissue and finally by breaking through to the surface. The rate at which these changes occur is heavily influenced by storage temperature, muscle type and degree of conditioning.[3]

The importance of colour as a marketing attribute of meat is well established, especially for self-service retailing. Customers, used to seeing bright-red meat prepared for sale, associate this colour with good eating quality, although there is little correlation between the two. The fact that consumers believe otherwise has been the dominant factor in fresh meat marketing; the attractive oxymyoglobin form of the pigment is taken to indicate freshness, while the appearance of brown metmyoglobin usually signals the end of display life for fresh meat. This usually occurs before the development of off-odours associated with microbiological spoilage.

The colour of frozen meat is initially dependent on freezing rate and the consequent size of ice crystals in the surface layer. Slow freezing produces large ice crystals with poor light-scattering properties which give meat a dark, translucent appearance; conversely, the small ice crystals formed with fast freezing scatter light to make the meat surface pale and opaque.

The principal deterioration in frozen meat during storage is photo-oxidation of the pigment. Under direct illumination frozen meat oxidises from the surface inwards, whereas the chilled meat oxidation begins in the subsurface layer and progresses towards the surface. As with chilled meat, loss of redness is detrimental to the marketing of frozen meat.

3. INFLUENCE OF PACKAGING

The primary function of a meat package, to contain meat and prevent its contamination, is easily accomplished with the range of plastic materials available today. However, packaging may also be used in a wider context to improve storage life, attract the customer and, more recently, to extend the period of attractive display.

These functions are achieved by creating a new environment for the meat, the most important change being modification of the gaseous atmosphere. The composition of this atmosphere determines the colour of meat and the nature of spoilage which develops. Gases dissolve in meat according to their partial pressures, and reactions in the meat may consume or produce gases. The package may not be a complete barrier to gas and the internal atmosphere is therefore an equilibrium of gas exchanges between meat, package and external atmosphere.

The gases of primary importance are carbon dioxide and oxygen. Carbon dioxide has bacteriostatic properties and oxygen is consumed by respiration of tissue and aerobic bacteria. Because of the importance of oxygen for meat colour, any restriction in supply to the muscle pigment will affect colour. All commercial packages of meat contain a certain amount of oxygen at some time, and packaging systems differ principally in the quantity available to the meat initially and thereafter during storage. The effects which environmental conditions have on meat microbiology have been reviewed in detail by Ingram[4] and Finne.[5]

3.1. Packaging Materials

The growth of meat packaging has been accompanied by the development of plastic properties which are appropriate for a variety of packaging systems. A plastic film derives its basic properties from the monomer unit of the polymer from which it is made. Monomers composed of carbon and hydrogen produce polymers such as polyethylene, which are good barriers to moisture but relatively permeable to gases. Inclusion of chlorine in the monomer unit greatly reduces gas permeability but may make the polymer film brittle. This disadvantage can be overcome by adding small quantities of other monomers such as acetals and acrylates which upset the regular polymer structure and make the film more flexible. Other compounds may be added to polymers to improve their handling on machines or enhance particular properties.

The method of production also affects its properties. Most meat-packaging films are thermoplastic and are extruded from the molten stage through dies. In many cases they are stretched to thinner gauges before cooling, and this imparts the ability to shrink when re-heated.

If the properties required of a packaging material cannot be satisfied by a single film, several with individual desirable properties may be combined to give a satisfactory composite. They may be combined by adhesive laminating two or more films together, or by co-extruding polymers together to form a single material. In some cases a composite is produced by coating a film with another polymer. Polyvinylidene chloride is commonly used in this way to produce materials with very good gas and moisture barrier properties.

The choice of films for packaging meat is largely determined by their moisture and gas permeabilities.[6,7] Since weight loss must be avoided, most films are moisture barriers. However, gas permeability is much more variable and is specific to individual polymers. Overwrapping films commonly have oxygen permeabilities above 10 000 $cm^3\ m^{-2}\ day^{-1}\ atm^{-1}$ O_2, while the composite materials used for vacuum packing may have oxygen permeabilities below 20 on the same scale.

The moisture and gas permeabilities of plastic films, laminates and composites used in meat packaging are shown in Table 1.

3.2. Vacuum Packing

Perhaps the greatest innovation in meat handling during the last 20 years has been the introduction of vacuum packing for distribution and storage of chilled beef. This system has grown most rapidly in the USA where the majority of beef is now transported as boxed vacuum packs. It is widely

TABLE 1

OXYGEN AND CARBON DIOXIDE PERMEABILITIES OF SOME PLASTIC FILMS AND COMPOSITES USED IN MEAT PACKAGING

	Permeability ($cm^3\ m^{-2}\ day^{-1}\ atm^{-1}$ gas)	
	Oxygen	*Carbon dioxide*
Polyethylene (low density)	8 500	44 000
Polyethylene (high density)	1 840	7 900
Polypropylene	3 000	7 900
Polyvinyl chloride (PVC)	4 200	17 000
Polyester (PET)	79	240
Polyvinylidene chloride (PVDC)	10	53
Nylon 6	240	1 600
Ionomer (Surlyn)	5 000	15 000
Ethylvinyl acetate (EVA)	12 000	38 000
Polystyrene	4 500	31 500
Cryovac BBI (2 mil)	20	80
Nylon/polyethylene (0·6/1·5 mil)	90	600

All permeabilities are for film thicknesses of 1 mil (25 μm) except where stated otherwise.

used in many countries for export and domestic markets and nearly half of UK beef is handled in this way.

There are clear advantages in transporting and storing beef as boneless joints instead of carcasses. Only about two-thirds of a beef carcass is usable meat, and refrigerated space is wasted by transporting and storing the remainder which is low value or waste. Vacuum packing also allows tenderness of beef to be improved by ageing without the evaporative weight loss incurred when carcasses are hung in the conventional way. The meat retailer has the further advantage of simpler butchery and greater versatility in stocking to satisfy consumer demand.

For vacuum packing, large boneless joints weighing between 3 and 15 kg are enclosed in flexible plastic containers, usually bags, to prevent moisture loss and exclude oxygen from the meat's surface. The plastic materials used for vacuum packing must have low moisture and gas permeabilities and be strong enough to hold heavy beef joints. Vacuum packing reduces the volume of air sealed in with the meat and restricts further entry of oxygen during storage. When a flexible pack is evacuated it collapses around the meat so that the pressure inside the package is seldom much less than atmospheric. The oxygen in the small volume of residual air is quickly consumed by meat respiration and within 2 days oxygen partial pressure at the surface of the meat will have dropped below 10 mm Hg. Most of the plastics used for vacuum packs have oxygen permeabilities below 100 $cm^3\ m^{-2}\ day^{-1}\ atm^{-1}\ O_2$ and replenishment of oxygen from outside is limited. At these very low partial pressures the penetration limit of oxygen is very near the surface and the thin brown layer of metmyoglobin which develops cannot conceal the underlying myoglobin. The visible colour of vacuum-packed beef is therefore purple. The surface layer of metmyoglobin may be as thin as 1 mm with very good barrier materials but as much as 2–3 mm with oxygen permeabilities approaching 200. This greater thickness of metmyoglobin may noticeably dull the purple colour of the underlying muscle. When vacuum packs are opened the purple myoglobin oxygenates and the colour becomes bright red. Again, this colour is clearer the thinner the oxidised surface layer.

As the large bulk of meat in a vacuum pack depletes oxygen, tissue respiration produces carbon dioxide which accumulates to more than 20% within a day or two of packing. CO_2 concentration remains above this level throughout storage and, together with the low oxygen level, produces an entirely different pattern of microbial growth from that which occurs in air. The major result is that the growth of pseudomonads responsible for spoilage of meat in air is limited and lactic acid bacteria become the dominant organisms. Unlike pseudomonads, these bacteria do not produce putrid odours and, with beef, can reach maximum numbers of about $10^8/cm^2$ after approximately 5 weeks at 1 °C without apparent spoilage.[1]

3.3. Vacuum Packaging Systems

Three basic methods are available for vacuum packing meat.

(*a*) *Cryovac Method*

This system, developed by W. R. Grace, is based on Cryovac machines using Cryovac bags. The currently used BBI Barrier bags are constructed from a triple co-extrusion of EVA/PVDC/EVA. They are tough, clear and, unlike most other vacuum bags, are heat-shrinkable. On early machines, evacuation was accomplished by inserting a metal tube into the neck of the bag to suck out the residual air, but the newer machines are fully automatic and accomplish more efficient evacuation by a double vacuum chamber. After exhaustion, a metal clip is applied around the twisted neck of the bag, and the bag heat-shrunk by dipping in water at 90 °C. The excess plastic shrinks to conform closely to the meat and produce a tight vacuum pack with an oxygen permeability low enough for good colour stability. The metal closure is claimed, justifiably so, to be as good as a heat-welded plastic seal.

In recent years some manufacturers have introduced heat-shrinkable bags which can be used in the same way as Cryovac bags. These include combinations of nylon and Surlyn which can also be sealed by metal clipping. In some cases, permeability to oxygen may be slightly higher than the Cryovac bag, but is generally still low enough to be effective for vacuum packing meat.

(*b*) *Chamber Method*

This technique was first used for vacuum packing bacon and other processed meats. The meat is put into a pre-formed bag which is then placed in an enclosed chamber which is evacuated. At a predetermined low pressure, heated jaws close and weld the mouth of the bag. When the chamber is opened to atmosphere, the bag collapses around the meat. The bags used in this type of system are made from laminates or co-extrusions of various polymer films. They usually include nylon or polyester as the outside layer to provide strength and good oxygen barrier properties and an inner layer of low-density polyethylene, Surlyn or EVA which are good moisture barriers and easy to heat-seal. The thicknesses of the component films can be varied; thicker nylon or polyester layers are better gas barriers, but are generally not as effective as composites containing PVDC.

Bags are produced in tubular form or with three heat-sealed sides and can be used on machines such as Swissvac, Multivac, Autovac, Supervac and Wrightvac, all based on the chamber principle. In most cases the chamber comprises a platform base with a hinged lid or hood which is closed down for evacuation.

Bags used in the chamber process are not heat-shrinkable and drip tends to accumulate in corners during storage. This may give the impression that this type of vacuum packing causes more drip than the Cryovac shrinkable bag, but several workers have found no difference in drip volume between the two packaging systems. Nevertheless, the accumulated drip is unsightly and secondary sealing has been introduced to overcome this problem. Vacuum packs containing Surlyn or EVA films as their inner layer are passed through a heating tunnel, and since they are very easily heat-sealed, the excess area around the meat is sealed. Not only is drip prevented from spreading into the corners, but the wider heat seal is also claimed to reduce the incidence of leaking bags.

(c) Thermoforming Method
This method, commonly used for vacuum packing smaller pieces of meat or processed meats, is now being adopted for packing larger joints. Fully automatic machines such as Multivac, Tiromat, Hammerle and Dixie-Union use moulds to pre-form deep trays from a base web of plastic. Meat is placed in the trays and an upper web of plastic is heat-sealed under vacuum to form a lid. The materials used for thermoforming are generally composites of nylon, polyester or polyvinyl chloride, possibly with PVDC coating and heat-sealing layers such as polyethylene, EVA or Surlyn.

This type of vacuum packing has a particular application with hot boned meat. A commercial operation in Denmark vacuum packs hot boned beef primal joints in appropriately shaped trays. This overcomes the difficulty encountered when inserting hot, pre-rigor meat into vacuum bags and provides a certain degree of moulding to the meat as it cools.

Some of the plastic materials used for thermoforming can withstand boiling water temperatures, and the meat can be cooked without unpacking. Used in this way, vacuum packing can be an aid to processing.

3.4. Vacuum Packing Beef

Vacuum-packed beef joints can be stored for at least 2 months at 1 °C before sour odours become noticeable. Such long-term storage is possible only with beef of normal ultimate pH, i.e. around 5·5. Occasionally beef carcasses are found with DFD (dark, firm, dry) muscle, sometimes described as 'dark-cutting'; such muscle has an ultimate pH of 6 or more. At this high pH, hydrogen sulphide-producing bacteria can grow, leading not only to development of putrid odours, but also to conversion of the muscle pigment on the meat surface to green sulphmyoglobin. It has been shown[8] that beef with ultimate pH only slightly above normal will keep better in bags with low gas permeability, but where the pH is above 6 (and the meat is dark in appearance), it will not keep much beyond 2–3 weeks at 1 °C in even the best barrier bags. Such beef should not, therefore, be vacuum packed.[8,9]

3.5. Vacuum Packing Lamb and Pork

Until fairly recently vacuum packing of chilled meat was confined to beef, but there is now interest in applying the technique to lamb and pork. Pork and lamb carcasses are too small for anything other than partial boning before packing and the presence of bone inevitably leads to a high incidence of punctured bags. This has restricted the application of vacuum packing to these species.

Studies at the Meat Research Institute[8] have shown that the storage stabilities of vacuum-packed lamb and pork are less than that of beef. Lamb loins, shoulders and legs in vacuum packs at 1 °C remained unspoiled for up to 6 weeks, but deteriorated rapidly during subsequent retail display at 5 °C. Because of this it was recommended that vacuum-packed storage should be no more than 4 weeks. Nevertheless, commercial shipments of vacuum-packed lamb, as joints or as whole carcasses cut and telescoped for ease of packing, have been sent successfully from New Zealand to the UK.[10] Careful attention to hygiene before packing and control of temperature during shipment (−1 °C rather than +1 °C) have improved storage stability and performance during retail display.

Trials at the Meat Research Institute have suggested a useful storage life of a little more than 2 weeks at 1 °C for vacuum-packed pork joints, although Smith *et al.*[11] claimed an acceptable storage life of 3–4 weeks for vacuum-packed loins at 2 °C. Scandinavian reports also suggest that pork joints have a longer storage life than in the MRI trials, and industry has therefore been encouraged to make limited use of vacuum packing for wholesale cuts of pork. The attraction is likely to be the convenience in butchery rather than savings in weight loss.

4. RETAIL PACKAGING

4.1. Overwrapping

Satisfying the consumer's preference for bright-red fresh meat is the prime objective of packaging for self-service retailing. It is generally achieved by maximum exposure of the meat surface to air, while preventing moisture or liquid loss. Meat is retailed on rigid or expanded plastic trays, overwrapped with clear plastic film held in place by heat-sealing or cling-folding. The overwrapping film, which may be a vinyl or polyethylene derivative, is generally very thin (15–25 μm) with a low moisture permeability and an oxygen permeability greater than 10 000 $cm^3\ m^{-2}\ day^{-1}\ atm^{-1}\ O_2$. It has long been considered that a plastic film must have an oxygen permeability of at least 5000 to allow sufficient oxygenation for an attractive red colour in meat.[12] This requirement is easily satisfied by the plastic films currently used for overwrapping trays of meat and the choice among them is likely to be based on suitability for use on machines, optical clarity, sealability and price rather than oxygen permeability.

In practice overwrapped meat is slightly less stable than carcass meat because the wrap prevents surface drying. Even so, with good refrigeration, pre-packed meat is acceptable microbiologically for up to a week. Retail life of chilled meat is limited by deterioration of the desirable bright-red colour; within 1–3 days display the layer of brown metmyoglobin which has formed 4 or 5 mm beneath the red surface begins to thicken and approaches the surface. Since the surface layer is translucent, the brown metmyoglobin becomes visible and the meat unsaleable. On extended display the metmyoglobin erupts through the surface and produces brown–green patches.

The influence of temperature on meat colour deterioration is well known; an increase from 0 to 5 °C doubles the rate of discoloration. It is clearly important to maintain the lowest possible chill cabinet temperature, but the influence of overwrapping on meat temperature is not always fully appreciated. Overwrapped trays provide very effective thermal insulation and the meat temperature is often higher than the surroundings. The lighting system in display cabinets may also heat the upper surfaces of packed meat by the 'greenhouse' effect. It is important therefore that meat is adequately cooled before packing, and low temperatures maintained not only in the cabinet but in the meat itself during storage and display. The stability of meat pigments to light is not yet fully understood, but there is some evidence that meat will retain its attractive colour longer if stored in the dark. A comprehensive review by Kropf[13] examines the effects of display conditions on meat colour.

4.2. Centralised Pre-packing

Although vacuum packing meat for wholesale distribution has removed the need for heavy butchery at the retail outlet, expensive space and labour are still required for preparing final retail packs. Industry is becoming increasingly aware of the economic advantages of distributing packed portions of meat directly to retail outlets from centrally located cutting and packing plants. As has already been pointed out, however, the display life of overwrapped portions of meat is only 1 or 2 days. This rapid colour deterioration has discouraged widespread introduction of centralised packing for chilled meat portions. A centralised operation requires packed meat with a display life approaching 1 week, and a number of packing systems now available can accomplish this. Methods for prolonging display life must pay strict attention to good refrigeration and hygiene, but they achieve their effect mainly by modifying the gas atmosphere around the meat. This is sometimes described as 'controlled atmosphere packing', but that term more correctly describes systems in which gas concentrations are continually adjusted during storage and transportation. The modified atmosphere systems discussed here involve sealing the meat in a gas-impermeable package with a specific gas mixture, the composition of which is changed to some extent by reactions with the meat.

Modified atmospheres can be used in two ways. In the first, the system is designed to exclude oxygen and keep the meat pigment in the purple myoglobin state until display, at which point it is allowed to oxygenate and bloom red in air. In the second method, the gas mixture includes oxygen so that maximum oxygenation at the time of packing produces the bright-red colour and retains it during storage and display.

4.3. Vacuum Packing

The simplest modified atmosphere package is a vacuum pack, and for a long time this has been suggested as the most effective way of extending the display life of chilled meat.[14] The successful use of vacuum packing for storage and transportation of beef primals indicates the potential advantage of extending the system to retail portions. Not only can a storage life of several weeks be achieved, but during this time the meat is improving in tenderness. Moreover, the close-fitting pack helps to conceal excessive drip and is ideally suited for home freezing.

Vacuum packing of retail cuts differs from vacuum packing of larger joints in one respect which is important for colour. With large joints, the ratio of residual air to meat is low enough to ensure quick depletion of oxygen, but with smaller meat portions the ratio can be higher and consequent metmyoglobin formation can discolour the surface of the meat. It is essential therefore, if retail cuts are vacuum packed, that they are packed with minimal residual air immediately after cutting.[15]

Claims have been made recently of improvements in colour and storage stability with more efficient evacuation.[16] This is accomplished with very low permeability films and by supplementary pumping units giving lower chamber pressures than conventional vacuum-packing machinery. Better storage performance with very low permeability films is also reported by Newton and Rigg.[17]

Colour remains the major handicap and, although the purple colour of vacuum-packed meat is accepted by the meat trader who knows that it will bloom red when exposed to air, it is believed that the consumer will reject meat which is purple. This is particularly unfortunate, since the purple colour of vacuum-packed meat is stable during prolonged display and this gives vacuum packing a considerable advantage over other methods of extended-life packing where colour gradually deteriorates from time of packing. It is difficult to see how the problem of consumer acceptance of vacuum-packed colour can be overcome without a massive education programme, but marketing fears may be overstated; vacuum-packed beef subprimals have been introduced in some US supermarkets alongside conventionally overwrapped meat. Sales of these subprimals have been successful, but retailers have not yet had the confidence to extend the practice to smaller portions.

Evidence from laboratory trials indicates that vacuum packing could be used commercially. Kropf[13] claims a shelf-life of 21 days at 2 °C for vacuum-packed beef muscles and similar results are reported by other workers.[18,19] Attempts to market vacuum-packed beef retail portions have been made in the USA[20] and in Sweden, where particular attention was given to consumer education on colour. These operations were successful but fairly localised, and have made no great impact on the industry. In the UK a limited range of vacuum-packed beef cuts has been marketed regionally, with some success, but here also total impact has been minimal. Vacuum-packed lamb legs, loins and shoulders have been imported to the UK from New Zealand for several years. During transportation, temperature is maintained at around −1 °C and the cuts can be successfully retailed later. The joints are generally removed from the packs before display and perhaps they should not be regarded as vacuum-packed retailed meat.

Colour is not as great a problem with pork since the difference in colour between oxygenated and reduced myoglobin is less pronounced. Vacuum-packed boneless pork joints are currently being retailed in the UK, Denmark, Sweden and the Netherlands. A shelf-life of 2 weeks at 1 °C is long enough for distribution of centrally packed retail joints over short distances.

Recent developments in vacuum skin packing (mentioned later in relation to frozen meat) and the appearance of new automatic machines suggest that this new technique may have a particular application in packing and retailing of vacuum-packed retail cuts of fresh meat.

4.4. Gas Packing

Vacuum packing has the inherent disadvantage that both package and meat are subjected to mechanical strain. Mechanical pressure on the meat may increase drip loss and if bone is present the pack may be ruptured. The strain on the packaging material can be alleviated by introducing another gas or mixture of gases after evacuation and before sealing.

(*a*) *Nitrogen*

As an inert gas, nitrogen is convenient for gas packing. If added after evacuation of air, the effect on meat is similar to that of vacuum packing, except that residual oxygen is diluted and metmyoglobin formation on the surface should be less pronounced than with a vacuum. However, although the purity of myoglobin at the surface is improved, the nitrogen also dilutes the carbon dioxide produced by tissue respiration, and the concentration required to inhibit growth of spoilage bacteria takes longer to accumulate.

Although nitrogen may be used instead of vacuum packing for sliced cured meats, it has had little commercial use with fresh meat. Nevertheless, several studies have confirmed that 100% nitrogen is as effective as vacuum for storing meat joints. Partmann *et al.*[21] obtained good colour retention in carcass parts during 7 days at 3°C. O'Keefe and Hood[22] showed that retail cuts packed in nitrogen lost less exudate than similar cuts in vacuum packs, and attributed this to mechanical pressure on meat when vacuum packed. Seideman[23] concluded that packing in 100% nitrogen had no advantage over vacuum packing other than reduced exudate and found no other differences with cuts stored under both conditions for up to 28 days.

(*b*) *Carbon Dioxide*

Gas atmospheres containing carbon dioxide have been used in meat storage for many years. Low concentrations (10–20%) of CO_2 were used more than 40 years ago in the refrigerated holds of ships carrying carcass meat from Australasia to Europe.[24] Since then, many workers have used higher concentrations to achieve effective extension of storage life over aerobically packed meat. Pure CO_2 has been used in some studies, but in most it has been mixed with nitrogen or air.[11,21,23,25–38] In general it was shown that storage life extension was similar to that achieved with vacuum packing; longer storage was obtained with higher concentrations of CO_2, and nitrogen, if present, played no active part.

In these studies oxygen was either completely absent or present in low concentrations, so that formation at the meat surface of oxymyoglobin was precluded or hindered, and that of metmyoglobin encouraged. These gas mixtures have therefore been of less interest in meat retailing than modified atmospheres which enhance the red colour during display.

Early use of CO_2 with meat brought adverse reports on colour if high concentrations were used.[39] Many of the previously cited workers found a similar effect, particularly where CO_2 hindered full oxygenation of the meat on subsequent exposure to air. It has not been observed at the Meat Research Institute, and it is difficult to see how the situation can be very different from that in vacuum packing, where CO_2 concentrations above 50% are common. Such vacuum-packed beef stored for several weeks oxygenates satisfactorily on exposure to air. The reported adverse effect on colour may have been caused by metmyoglobin formed from residual oxygen left in the package after evacuation.[40] Unless scavenging is used, it is difficult to exclude oxygen completely and its presence would be particularly harmful in retail packs since it discolours the display surface.

Most systems using CO_2 with N_2 are two-stage packing but, in a method patented by Hirsch *et al.*,[41] anaerobic storage is followed by aerobic display in the same package. A meat portion is sealed with a CO_2/N_2 mixture in a gas-impermeable, lidded tray. During storage oxygen is excluded, but for display the impermeable outer film of the laminate forming the lid of the tray is peeled off to expose a highly oxygen-permeable film. Oxygenation of the meat is complete in 30 min and the red colour is claimed to be stable for up to 7 days at 2 °C. Although the concept was publicised several years ago and has been used in Scandinavia, it does not appear to have been introduced in the UK, possibly because complete removal of oxygen is difficult and excessive drip accumulates in the package during storage.

While the application of high CO_2 concentrations with beef and lamb may be discouraged by adverse effects on colour, they may be more useful with pork. Partmann *et al.*,[31] Seideman *et al.*[34] and Enfors *et al.*[35] have demonstrated that pork cuts in 100 % CO_2 have a microbiological storage life of 5–6 weeks at 4 °C. This is better than trials at the Meat Research Institute where storage life of vacuum-packed pork was no more than 2 weeks at 1 °C. Continuing Scandinavian interest in the use of high CO_2 concentrations for gas flushing as well as for gas storage, together with claims of even longer storage life for pork, suggest the feasibility of retailing chilled pork cuts in CO_2 packages.

(*c*) *Oxygen and Carbon Dioxide*

Although the relationship between high oxygen concentration and oxymyoglobin formation is well known, there are few early reports on the use of such concentrations with meat. Fellers *et al.*[42] obtained attractive redness with 33 % O_2, confirming observations made by Bratzler[43] using higher concentrations. Zimmerman and Snyder[44] attempted to use high-pressure oxygen treatment to inhibit respiration and prevent formation of metmyoglobin. Beef slices were treated with O_2 at more than 5 atm for 12 h, to oxygenate them completely, but when they were subsequently wrapped in impermeable film, respiration proceeded normally and oxygen depletion at the surface led to formation of metmyoglobin. Similar slices held in high-pressure oxygen for 12 days were still oxygenated at the end of that period, but off-odours had developed, possibly from lipid oxidation or aerobic spoilage.

In an experiment carried out at the Meat Research Institute[3] fresh beef, overwrapped in permeable film, was transported in oxygen to a retail outlet where its colour was compared during display with similar meat which had been transported in air. Colour deterioration was successfully delayed by holding in oxygen during the one-day transport and storage period, the process being most effective with muscles prone to rapid discoloration. There is little doubt that high concentrations of oxygen hold the bright-red colour, but microbiological stability under these conditions is unsatisfactory, and technical problems can arise when oxygen is used at high pressures.

Although many studies combined CO_2 with O_2 for packing meat, the O_2 concentrations were generally less than that in air and the effect on colour was detrimental. Little interest was shown in utilising the bacteriostatic properties of CO_2 and the colour-enhancing properties of high oxygen concentrations until Brody[45] reported that such mixtures gave good colour and keeping quality. Naumann *et al.*[46] described experiments in which mixtures of 15% CO_2 and 85% O_2 improved the packaged life of beef steaks held at -1 °C.

Commercial interest in the use of CO_2/O_2 mixtures for packing meat was signalled in Europe by two patents in 1970. They covered the use of various combinations of CO_2 and O_2 for packing meat in low-permeability, deep-drawn, lidded containers to obtain a useful display life of 1–2 weeks. In the first,[47] air in the meat container was replaced with gas mixtures of 50–70% O_2 with 50–30% CO_2. Best results were obtained with 2:1 ratios of O_2:CO_2, but shelf-life was conservatively claimed at 'more than 3 days at refrigeration temperatures near freezing'.

The other patent[48] is more interesting since it deals with O_2 concentrations above 70% combined with at least 10% CO_2. The patent specification evaluated in detail the microbiological and colour shelf-life achieved with the preferred gas mixtures, and with others found to be less effective. The results in Tables 2 and 3 summarise the requirements for successfully gas packing meat. Table 2 shows the colour-enhancing effect of high O_2 concentration in mixtures 1 and 3 and the bacteriostatic effect of CO_2 in 1, 2 and 4. Both advantages are achieved with the 80% O_2 + 20% CO_2 mixture. Assessment of colour relied on subjective judgement of the change from red to brown and these results should only be regarded as comparisons between treatments.

Table 3 shows the decrease in colour purity during storage. The colour

TABLE 2

MICROBIOLOGICAL AND COLOUR ACCEPTABILITY OF BEEF CUTS PACKED IN GAS MIXTURES AT 4 °C[48]

Gas mixtures	*Total viable count (bacteria/g)*			*Colour*		
	5 days	*10 days*	*15 days*	*5 days*	*10 days*	*15 days*
1. $80\% O_2 + 20\% CO_2$	$<10^7$	$<10^7$	$<10^7$	Red	Red	Red
2. $20\% O_2 + 20\% CO_2 + 60\% N_2$	$<10^7$	$<10^7$	$<10^7$	Red	Brown	Brown
3. $80\% O_2 + 20\% N_2$	$>10^7$	$>10^7$	$>10^7$	Red	Red	Red
4. $100\% CO_2$	$<10^7$	$<10^7$	$<10^7$	Brown	Brown	Brown
5. Air	$<10^7$	$>10^7$	$>10^7$	Red	Brown	Brown

Note: In this comparison $<10^7$ bacteria/g was judged 'acceptable' and $>10^7$ 'unacceptable'. Colour was judged by a trained panel, red being 'acceptable' and brown 'unacceptable' from the point of view of customer preferences.

change from acceptable (red) to unacceptable (brown) was judged to have occurred when colour purity fell below 34 %. The results demonstrate that at least 60 % O_2 is required to prolong colour for 9 days and the patent claims that a mixture of 80 % O_2 + 20 % CO_2 keeps meat red for up to 15 days at 4 °C, during which time microbial spoilage is kept to an acceptable level.

TABLE 3

COLOUR SCORES FOR BEEF CUTS PACKED IN DIFFERENT PROPORTIONS OF O_2 AND CO_2 AT 4 °C[48]

Days at 4 °C	*Colour purity (%)*					
O_2/CO_2:	*90/10*	*80/20*	*60/40*	*40/60*	*20/80*	*10/90*
0	42·5	42·5	42·5	42·5	42·5	42·5
6	36·3	36·0	37·4	36·0	30·8	27·8
9	35·5	35·0	34·0	28·0	25·4	27·2
13	32·5	28·0	25·5	25·5	25·6	25·3
Days to reach critical purity of 34 %	11	10	9	7	<6	<6

Scores are for 'colour purity'[49] measured using a 'Colocord' colorimeter (Joyce Loebel Ltd).

Since containers for gas packing are gas barriers, the internal atmosphere will be modified by the meat during storage. The relative volumes of gas and meat are therefore important in determining the progress of the changes in concentration of gases during storage. Studies carried out by Taylor[50] and Taylor and MacDougall[51] followed the changes in gas volume and composition when beef was stored in completely impermeable containers with mixtures of CO_2 and relatively high concentrations of O_2. Figure 1 shows that an initial O_2 concentration of 80% was depleted to approximately 65% within 2 days, where headspace volume was roughly equal to meat volume as might be obtained in a commercial package. During this time the CO_2 level changed only slightly since the gas dissolving in the meat was balanced by CO_2 from tissue respiration. Oxygen level during storage determines the effectiveness of colour retention. The penetration depth with 80% concentration of the surface oxygen is twice as deep as with air. The metmyoglobin layer formed at the limit of penetration is so far from the surface that it is concealed for several days. The oxymyoglobin pigment is, however, still subject to slow

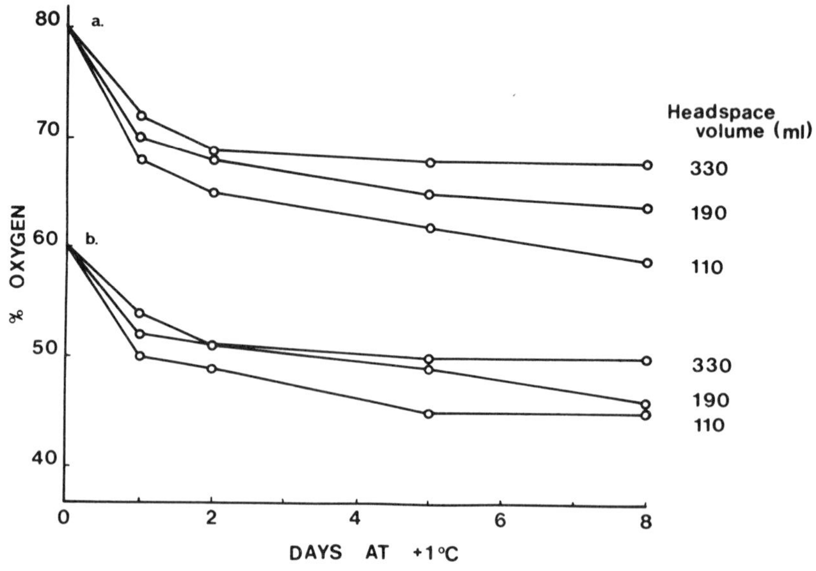

FIG. 1. Changes in O_2 concentration in containers of 130 g meat with different initial headspace volumes of gas mixtures: (a) 80% O_2 + 20% CO_2; and (b) 60% O_2 + 20% CO_2 + 20% N_2.

TABLE 4

THICKNESS OF OXYGENATED LAYER IN BEEF SAMPLES STORED IN DIFFERENT MIXTURES OF O_2 AND CO_2 [51]

Gas mixture	*Thickness of oxygenated layer (mm)*				
	1 day	*2 days*	*5 days*	*8 days*	*12 days*
$80\%\ O_2 + 20\%\ CO_2$	11	11	14	14	*
$60\%\ O_2 + 20\%\ CO_2 + 20\%\ N_2$	9	9	8	*	*
$40\%\ O_2 + 20\%\ CO_2 + 40\%\ N_2$	8	7	*	*	*

* Observations where surface layer was no longer bright red.

oxidation to metmyoglobin and with time the initial redness becomes diluted with brown. This change occurs gradually and uniformly throughout the oxygenated layer and, in the early stages, is difficult to detect except by direct comparison.

Table 4 shows the depth of oxygenation with 80%, 60% and 40% concentrations and indicates the time by which surface colour is no longer considered red. Other workers have also concluded that mixtures of 75–85% O_2 with 25–15% CO_2 are most effective.[28,29,32,33,36,52,53] However, there have also been reports of off-odours and rancidity in meats stored in high O_2 concentrations.[34,54]

(*d*) *Carbon Monoxide*

The possibility of accelerated fat oxidation with high oxygen concentrations has led to some interest in the use of carbon monoxide. Matamoros and Rama[55] showed that meat packed in CO developed a bright-red colour due to formation of the pigment, carboxymyoglobin, which was stable during store. El Badawi *et al.*,[56] Clark *et al.*, [57] and Gee and Brown[58] obtained similar results, but only when the CO was present throughout storage. Samples of meat treated with CO gas, even at 99% concentration, lost their red colour as quickly as untreated samples when subsequently exposed to air. Wolfe[38] stated that although CO retarded the formation of metmyoglobin and avoided fat oxidation, its use with meat was not approved by regulatory authorities. There are few arguments against these views, although Tappel *et al.*,[59] Clark *et al.*[57] and Watts[60] have pointed out that the health hazard of CO may have been exaggerated since, if used in low concentrations, the body's exposure to it would be minimal.

(*e*) *Commercial Modified Atmosphere Packing*

Several European countries are currently using mixtures of 60–80% O_2 and 40–20% CO_2 to prolong the bright-red colour in chilled meat for at least a week. Various machines are available for commercial-scale packing, many of them thermoforming machines modified to inject gas mixtures before sealing. In Europe, Multivac, Tiromat, Hammerle, Swissvac and Dixie Union machines are used to form trays from a bottom web of film and, after evacuation and flushing, seal lids over the trays. Rose Forgrove and Hayssen machines use a different principle, forming around trays of meat on a conveyor belt a continuous tube into which a long lance injects the gas mixture before sealing around each tray. Typical plastic films used for thermoforming and gas packing are PVC/polyethylene for the rigid base and polyester/polyethylene for the lid. The alternative 'pillow-pack' type of system uses rigid pre-formed PVC trays which are enclosed in pouches made from PVDC-coated polyester/polyethylene or nylon/polyethylene.

Careful package design and system operation are essential to avoid problems which can arise with gas packing of meat. Drip, which can be a problem in overwrapped trays of meat held for only 1 or 2 days, will be even more evident in packs designed for longer display periods. Trays are therefore designed with patterned bases to disperse the liquid, but in most cases absorbent pads are also included. Excessive drip may prove a greater limitation than colour deterioration to acceptance of gas-packed meat by the consumer.

Another problem caused by moisture is condensation on the inside of the container lid. If excessive, this can defeat the whole purpose of gas packing since it obscures the meat, but it can be avoided by minimising temperature fluctuations in the display cabinet and by an anti-fog coating on the inner surface of the lid.

Automatic packing and sealing equipment is always liable to produce faulty packs and this is a particularly difficult problem in gas packing. Most monitoring systems rely on periodic sampling of finished packs to assess the incidence of leakers, since on-line identification of the faulty packages is difficult. The problem is even more acute with gas mixtures containing high levels of oxygen. With these, identification of leakers cannot be based on colour since, in the early period after packing, all the meat is red. By the time colour deterioration shows that a pack contains air and not a high O_2 concentration, the meat may be unsaleable.

The meat surface is oxygenated only where oxygen can get easy access; contact between meat and container or lid limits access and less attractive patches of brown metmyoglobin are produced. Trays are designed to minimise contact with the underside of the meat and be deep enough to avoid contact with the lid. The excessive size of tray, often twice as deep as the meat, adds considerably to packaging costs.

Machines used with high concentrations of oxygen may have to be modified to avoid possible fire or explosion hazards; special oils and greases are required for vacuum pumps, and electrical features such as heat-sealing mechanisms must be made safe.

Modified atmosphere packing of fresh meat requires strict attention not only to the packing operation but also to the temperature of the meat and its hygienic condition.[61] A Danish company, in a system using a high concentration of O_2 with CO_2, specifies hygienic slaughter, chilling without delay to 2 °C, no more than 4 days conditioning before portioning, cutting room temperature at or below 5 °C, packing immediately after cutting, and storage or display at or below 2 °C. Most importantly, the temperature of the meat must not exceed 2 °C at the time of packing, since the gas space around the meat makes subsequent cooling difficult. This specification satisfies Danish veterinary legislation allowing fresh meat to be retailed up to a week after packing.

Provided hygiene and refrigeration are good, modified atmosphere packing in high concentrations of oxygen with carbon dioxide would appear to offer nearly everything required for a centralised meat pre-packing operation. The principal disadvantages are high package cost (perhaps 5 or 6 times as much as overwrapping), accumulation of drip and, because of poor heat transfer and accelerated rancidity in high O_2 concentrations, unsuitability for frozen storage. The technique has another disadvantage compared with packing in the absence of oxygen: the duration of the attractive red meat colour is limited to a week and any ageing which takes place in the pack to improve tenderness is at the expense of colour. A short ageing period of a few days before packing, although adding to production costs, may therefore be advisable to ensure that the meat is tender at retail.

Despite these disadvantages, and principally because the meat retains its familiar appearance, modified atmosphere packing with high concentrations of O_2 and CO_2 is the most likely alternative to conventional overwrapping, short-life systems for fresh meat.

(f) Frozen Meat Packing

Modified atmosphere packing may encourage the introduction of centralised packing of fresh meat, but eventually the high cost required for only a moderate extension of useful display life may limit its widespread acceptance. As a viable alternative the meat industry should examine more closely the greater potential advantages in terms of cost, convenience and long-term stability of marketing fresh meat in the frozen state. Retail packaging of frozen meat is comprehensively reviewed by Kropf,[62] and this chapter will discuss only some of the recent developments which have improved packaging of frozen meat to a level at which it can compete with packaged chilled meat. Frozen meat is stored and displayed between -10 and $-30\,°C$, at which temperatures microbiological growth is arrested. The changes in meat most influenced by packaging are therefore those associated with appearance, and this is where most advances have been made.

The first requirement of a frozen meat package is to prevent moisture loss or transfer within the package. The package must be a moisture barrier and must also be in close contact with the meat, since space between package and meat allows accumulation of frost. The package must also be able to withstand puncturing from the sharp edges of frozen meat.

Colour is the most important feature of frozen meat appearance, and if a bright-red colour is required, it must be produced by oxygenation of the meat surface before freezing. After freezing the red colour is best maintained in a package which is relatively permeable to oxygen rather than in one which is a barrier. Several workers support this view,[63-65] although the minimum oxygen permeability of $7750\,cm^3\ m^{-2}\ day^{-1}\ atm^{-1}\ O_2$ stated by Price and Schweigert[64] is high compared with other studies.

Several packaging systems are now capable of satisfying these requirements for frozen meat. Some are vacuum-packing machines modified to allow heat shrinking after sealing, but the development which offers the most interesting prospect is vacuum skin packing. This technique was first introduced to use an ionomer film, Surlyn, which on heating softens to such an extent that it can be draped over sharp objects without puncturing. The object of vacuum skin packing is to sandwich a portion of frozen meat between a heat-softened top web of Surlyn and a lower web of the same film. Evacuation before sealing moulds the softened top web to the shape of the meat. There is no space between meat and package and frosting cannot develop to mask the attractive appearance. Some machines, such as the Bivac (Du Pont) and the Dixie Union, can handle either already frozen cuts or chilled cuts for freezing afterwards, but others, including the RF 400 (Rose Forgrove), can pack only meat which is already frozen. In this case the final radiant heat-sealing operation glazes the surface of the frozen meat portion and produces an appearance similar to that of chilled meat.

Surlyn film has a moderately high permeability and will keep the bright-red colour for at least a year if the meat is stored in the dark at −20 °C. However, when exposed to light, as in display, the red colour begins to darken after about a week. This is caused by light-activated oxidation of the pigment in the meat surface and is inevitable in meat which has been frozen in the bright-red state and subsequently exposed to light. The bright-red display period is nevertheless more than adequate for a centralised packing operation, with cuts of meat retailed in the frozen state replacing over-wrapped chilled meat; the frozen meat would be unlikely to be on display for longer than 1 or 2 days.

5. CONCLUSION

There is no doubt that techniques already exist for large-scale centralised packaging of meat, and increasing labour costs may create the right climate for their introduction. The greatest problems may, however, still lie in marketing, either in selling meat which is not bright red, or dealing with consumer prejudice against frozen meat, but indications now suggest that these obstacles can be overcome.

6. REFERENCES

1. Dainty, R. H., Shaw, B. G. and Roberts, T. A. (1983). In *Food microbiology: advances and prospects* (Roberts, T. A. & Skinner, F. A. (Eds)), Soc. Appl. Bact. Symp. Series No. 11, p. 151. Academic Press, London.
2. Brooks, J. (1938). *Fd Res.*, **3**, 75.
3. MacDougall, D. B. and Taylor, A. A. (1975). *J. Fd Technol.*, **10**, 339.
4. Ingram, M. (1962). *J. Appl. Bact.*, **25**, 259.
5. Finne, G. (1982). *Fd Technol.*, **2**, 128.
6. Eustace, I. J. (1981). *J. Fd Technol.*, **16**(1), 73.
7. Rigg, W. J. (1979). *J. Fd Technol.*, **14**(2), 149.
8. Taylor, A. A. and Shaw, B. G. (1977). *J. Fd Technol.*, **12**, 515.
9. Nicol, D. J., Shaw, M. K. and Ledward, D. A. (1970). *Appl. Microbiol.*, **19**, 937.
10. Bruce, J. H. and Roberts, C. A. (1982). *Refrig. Sci. Technol.*, p. 133.
11. Smith, G. C., Seideman, S. C., Savell, J. W., Dill, C. W. and Vanderzant, C. (1983). *J. Fd Prot.*, **46**(1), 47.
12. Sacharow, S. (1974). *Food Eng.* (March), p. 87.
13. Kropf, D. H. (1980). *Proc. 33rd Ann. Recip. Meat Conf.*, Amer. Meat Sci. Assoc., Chicago, p. 15.

14. Dean, R. W. and Ball, C. Q. (1958). *Agric. Fd Chem.*, **6**, 468.
15. Seideman, S. C., Carpenter, Z. L., Smith, G. C. and Hoke, K. E. (1976). *J. Fd Sci.*, **41**, 732.
16. Bohnsack, U. and Hopke, H. U. (1982). *Fleischwirtsch.*, **62**(1), 38.
17. Newton, K. G. and Rigg, W. J. (1979). *J. Appl. Bact.*, **47**(3), 433.
18. Griffin, D. B., Savell, J. W., Smith, G. C., Vanderzant, C., Terrell, R. N., Lind, K. D. and Galloway, D. E. (1982). *J. Fd Sci.*, **47**(4), 1059.
19. Vanderzant, C., Hanna, M. O., Ehlers, J. G., Savell, J. W., Smith, G. C., Griffin, D. B., Terrel, R. N., Lund, K. D. and Galloway, D. E. (1982). *J. Fd Sci.*, **47**(4), 1070.
20. Ernst, L. J. (1980). *Proc. 33rd Ann. Recip. Meat Conf.*, Amer. Meat Sci. Assoc., Chicago. p. 37.
21. Partmann, W., Frank, H. K. and Gutschmidt, J. (1970). *Fleischwirtsch.*, **9**, 1205.
22. O'Keefe, M. and Hood, D. E. (1981). *Meat Sci.*, **5**(4), 267.
23. Seideman, S. C., Smith, G. C., Carpenter, Z. L., Dutson, T. R. and Dill, C. W. (1979). *J. Fd. Sci.* **44**, 1036.
24. Haines, R. B. (1933). *J. Soc. Chem. Ind. Lond.*, **52**, 13T.
25. Kraft, A. A. and Ayres, J. C. (1952). *Fd Technol.*, **1**, 8.
26. Pohja, M. S., Alivaara, A. and Sorsavirta, O. (1967). *Proc. 13th Europ. Meeting Meat Res. Workers*, Rotterdam.
27. Ledward, D. A. (1970). *J. Fd Sci.*, **35**, 33.
28. O'Keefe. M., Hood, D. E. and Harrington, M. G. (1975). *Proc. 21st Europ. Meeting Meat Res. Workers*, Berne.
29. Huffman, D. L., Davis, K. A., Marple, D. N. and McGuire, J. A. (1975). *J. Fd Sci.*, **40**, 1229.
30. Partmann, W. and Bohling, H. (1977). FRG Patent Application 2,538,681.
31. Partmann, W., Bomar, M. T., Hajek, M., Bohung, H. and Schlaszus, H. (1977). *Fleischwirtsch.*, **57**(7), 1311.
32. Ordonez, J. A. and Ledward, D. A. (1977). *Meat Sci.*, **1**, 41.
33. Seideman, S. C., Carpenter, Z. L., Smith, G. C., Dill, C. W. and Vanderzant, C. (1979). *J. Fd Prot.*, **42**, 233.
34. Seideman, S. C., Carpenter, Z. L., Smith, G. C., Dill, C. W. and Vanderzant, C. (1979). *J. Fd Prot.*, **42**, 317.
35. Enfors, S. O., Molin, G. and Ternstroe, A. (1979). *J. Appl. Bact.*, **47**, 197.
36. Partmann, W. (1980). *Schlach. Vermarkt.*, **80**(10), 305.
37. Seideman, S. C., Vanderzant, C., Smith, G. C., Dill, C. W. and Carpenter, Z. L. (1980). *J. Fd Prot.*, **43**, 252.
38. Wolfe, S. K. (1980). *Fd Technol.*, **34**, 55.
39. Brooks, J. (1933). *J. Soc. Chem. Ind. Lond.*, **52**, 17T.
40. Hess, E., Ruosch, W. and Breer, C. (1980). *Fleischwirtsch.*, **60**(8), 1513.
41. Hirsch, A., Spiegel, F. X. and Ramsbottom, J. M. (1977). US Patent 4,055,672.
42. Fellers, D. A., Wahba, I. J., Caldano, J. C. and Ball, C. O. (1963). *Fd Technol.*, **17**, 1175.
43. Bratzler, L. J. (1955). *Proc. 7th Meat Ind. Res. Conf.*, Amer. Meat Inst., Washington, DC.
44. Zimmerman, G. L. and Snyder, H. E. (1969). *J. Fd Sci.*, **34**, 258.

45. BRODY, A. L. (1970). *Modern Packaging* (Jan.), p. 81.
46. NAUMANN, H. D., GONZALES, R. R. and YEH, L. C. (1971). *Proc. 17th Europ. Meeting Meat Res. Workers*, Bristol.
47. SCHWEISFURTH UND KALLE AG (1970). British Patent 1,186,978.
48. GEORGALA, D. L. and DAVIDSON, C. M. (1970). British Patent 1,199,998.
49. WRIGHT, W. D. (1964). *Measurement of colour*, 3rd edn. Hilger & Watts, London.
50. TAYLOR, A. A. (1971). *Proc. 17th Europ. Meeting Meat Res. Workers*, Bristol.
51. TAYLOR, A. A. and MACDOUGALL, D. B. (1973). *J. Fd Technol.*, **8**(4), 453.
52. CLARK, D. S. and LENTZ, C. P. (1973). *Can. Inst. Fd Sci. Technol. J.*, **6**, 194.
53. SAVELL, J. W., SMITH, G. C., HANNA, M. O. and VANDERZANT, C. (1981). *J. Fd Prot.*, **44**(12), 923.
54. LOPEZ-LORENZO, P., HERNANDEZ, P., SANZ-PEREZ, B. and ORDONEZ, J. A. (1980). *Meat Sci.*, **4**, 89.
55. MATAMOROS, E. G. and RAMA, A. M. (1973). *Proc. 19th Europ. Meeting Meat Res. Workers*, Paris.
56. EL-BADAWI, A. A., CAIN, R. F., SAMUELS, C. E. and ANGLEMEIER, A. F. (1964). *Fd Technol.*, **18**, 159.
57. CLARK, D. S., LENTZ, C. P. and ROTH, L. A. (1976). *Can. Inst. Fd Sci. Technol. J.*, **11**, 363.
58. GEE, D. L. and BROWN, W. D. (1978). *J. Agric. Fd Chem.*, **26**, 273.
59. TAPPEL, A. L., HARPER, J. C., REGIER, L. W., CONROY, A. and STEWART, G. F. (1954). *US Quartermaster Corps. Progr. Rept*, **7**, 21.
60. WATTS, B. M. (1954). *Adv. Fd Res.*, **5**, 1.
61. ZEUTHEN, P. and KNUDSEN, I. M. (1981). *Proc. 27th Europ. Meeting Meat Res. Workers*, Vienna.
62. KROPF, D. H. (1983). *Proc. Meat Sci. Tech. Symp.*, Lincoln, Nebraska, 1982 (Franklin, K. and Cross, H. R. (Eds)), National Livestock and Meat Board, Chicago.
63. SANDBERG, M. L. (1970). 'Effects of freezing system, freezing rate, film permeability and display case temperature and colour stability and weight loss of frozen beef longissimus and psoas major muscles', MS thesis, Kansas State University, Manhattan.
64. PRICE, J. F. and SCHWEIGERT, B. S. (1971). *The science of meat and meat products*, W. H. Freeman & Co., San Francisco.
65. HUNT, M. C., SMITH, R. A., KROPF, D. H. and TUMA, H. J. (1975). *J. Fd Sci.*, **40**, 637.

Chapter 5

IRRADIATION PRESERVATION OF MEAT AND MEAT PRODUCTS

P. S. Elias

Federal Research Centre for Nutrition,
Karlsruhe, Federal Republic of Germany

1. INTRODUCTION

Ionising radiation has been proposed for the preservation of food since the early 1930s. Indeed the first patent was issued in 1930 for the use of X-rays to prolong the storage life of food.[1] Because commercial utilisation of this technology required powerful radiation sources, little progress was made until appropriate γ-ray sources from radioactive isotopes and electron-beam generators of adequate power became available. The potential application of electron-accelerators for food processing was first described in the USA.[2] Since that time an increasing number of research programmes was initiated in many countries of the world. By the early 1970s about 55 countries had such programmes on food irradiation;[3] yet acceptance of the new technology by food industry, by governmental authorities and by the general public developed only very slowly and is still spurious today.

The first country to permit the marketing of an irradiated food (viz. potatoes irradiated for inhibition of sprouting), was the USSR in 1958. This was followed in the US in 1963 by the clearance of wheat and wheat flour irradiated for insect disinfestation and of canned bacon irradiated for long-term storage without refrigeration. Subsequent progress has been slow because too few foodstuffs were allowed to be irradiated and export barriers remained because of the legal prohibition of the treatment of foods by irradiation in most countries.

The irradiation of meat and meat products has remained largely a domain for research into the application of large doses of radiation for the sterilisation of meat and meat products, as an alternative to canning and heat pasteurisation, without the pursuit of any practical commercial objectives. Only recently has the design of irradiation facilities reached the stage in which low dose application to large volumes of meat for elimination of certain pathogens and parasites has become a practical feasibility.

2. THE EFFECTS OF IRRADIATION ON FOODSTUFFS

2.1. General Aspects

The complex composition of foodstuffs leads to the formation of a large number of reaction products as a result of irradiation. The radiolytic compounds and radicals produced by irradiation have been studied in complex foodstuffs as well as in model systems containing only pure food constituents. Generally, only quantitative rather than qualitative differences have been found in the radiolytic products identified. Dilute aqueous solutions of amino acids, sugars, vitamins and solutions of lipids are more sensitive to radiation than the same compounds as constituents of foodstuffs.[4] The increased radiation resistance of the latter is related to the higher concentration of constituents, to the presence of radiation-protective substances, and to the distribution of radiation damage among the large number of constituents of a foodstuff.

The destruction of micro-organisms or insects by radiation processing is the result of some chemical effect on some essential cellular constituent. Yet the severe biological effect is not paralleled by a similarly extensive chemical change in the food. The reason for this apparent discrepancy is the part the damaged cell constituent plays in the life of the irradiated organism. Because the radiation sensitivity of macromolecules is approximately proportional to their molecular weight, nucleic acids are more severely damaged than enzymes or individual amino acids in cells exposed to a given radiation dose. Even though the actual radiation damage may be small and similar in extent in each case, the effect on DNA will be lethal, the damage to enzymes barely recognisable, and the loss of amino acids undetectable.[4]

2.2 Factors Influencing Radiation Effects

Substrate concentration affects the end result of irradiation. A given radiation dose causes the formation of a certain number of primary radiolysis products. In aqueous solutions, these are mainly $H^{\cdot}$, $OH^{\cdot}$ and e_{aq}^{-}. These free radicals subsequently react with the solute so that the absolute amount of solute destroyed remains the same whatever its concentration, except that at high concentrations the direct effects of radiation on the solute predominate over indirect effects resulting from interaction with the free radicals derived from the solvent. The chemistry of deoxygenated systems is governed by reducing radicals and H atoms. Hence dimerisation and disproportionation of free radicals occur preferentially. In oxygenated systems the solute-derived free radicals react with oxygen to form peroxy radicals, oxidised products and superoxide radicals. Peroxy radicals then yield hydroperoxides which are largely unstable. In aqueous systems, frozen at $-180\,°C$, free radicals are immobile and survive for days after irradiation. They disappear however on thawing. The absence of water in dehydrated systems explains their lower reactivity and tolerance of higher irradiation doses.[72]

Complex foods also contain substances which act as protectors against radiation damage by taking up radiation energy or by reacting preferentially with the primary radiolysis products. This competition effect depends on the concentration of the protective substance and on the rate constants of the reactions with the primary radiolysis products of water. Examples of protectors are ascorbic acid, albumin, cysteine, methionine, cystine, thymine, maleic acid, orotic acid, etc. The presence of radiation protectors in food reduces the overall radiation damage by removal of the free radicals produced in water, which are the major reactants with enzymes and proteins. It should also be remembered that the radiolysis products formed in foods may differ from those found in irradiated solutions of pure food components, and their concentrations depend, of course, on the amounts of protector substances present in foods. Moreover, the same radiolysis product may be formed from chemically related substances, e.g. hydrogen sulphide can derive from cysteine, methionine or glutathione. Because many compounds in food can act as radical scavengers, it is impossible to predict which protecting substance will be affected in a given food on irradiation.

The spatial arrangement of molecules also affects the overall radiation sensitivity of a substance. Random coil conformation may be less sensitive than helix conformation. The enzyme–substrate complex is more resistant than the enzyme alone. Similarly the physical state of fat during irradiation is known to affect the quantities of radiolysis products formed.[5]

There is no precise knowledge of the contribution of spatial and structural effects on the radiation stability of foodstuffs. For meat more severe damage is noted in collagen than in globular proteins because of the highly organised triple helical conformation of collagen, which is more easily disturbed by disruption of hydrogen bonds.[6]

Moderation of radiation effects is also more obvious in multi-component foodstuffs such as meat with its aqueous and lipid phases than in single-component foods such as crystalline monosaccharides.

2.3. Reduction of Radiation Effects on Food Quality

Radiation-induced changes may alter taste and odour of foodstuffs in an undesirable manner. Quantitatively small alterations may suffice to produce noticeable changes in flavour or colour, especially in protein-rich foods subjected to high-dose treatment. Additives have been used to reduce or prevent these undesirable changes. Examples are 0·5 % of ascorbic acid in chopped cured meat, gallic acid and other polyphenols in oils and fats, tocopherol, alone or with ascorbic acid, in certain cured meats. However, some experimenters using ascorbic acid addition to ground meat and sausages have noted no beneficial effects on off-flavours.[7] Some doubt exists, therefore, whether radical scavenging has been the only mechanism of protection against the organoleptic effects of irradiation. Some additives, e.g. BHT and pyrophosphate have been used to improve the post-irradiation storage behaviour of meat.[8]

Additives have little protective effect on the vitamin content of irradiated foods. Thus 0·1 % glutathione or sodium nitrite reduced the thiamine loss in irradiated minced beef by about 10–20 %. The addition of 0·5–1·5 % sodium chloride and 0·3–0·5 % sodium tripolyphosphate sometimes combined with nitrite, ascorbate and spices appeared to preserve the taste and texture of irradiated bacon, beef, chicken, corned beef, ham, pork, pork sausage and other foods.[9] However, in general, the use of free radical scavengers has not become a widespread practice in food irradiation.[4]

Irradiation at low temperatures of −40 °C or thereabouts has been suggested as a measure to avoid undesirable chemical reactions during irradiation, based on the assumption of lowered reactivity and reduced diffusion rates of free radicals. Experimental evidence pointed to the need for a minimum temperature of −20 °C for noticeable protection of quality in sterilisation by irradiation (radappertisation). The advantages of low temperature were reflected in improved flavour, reduced production of hydrogen sulphide from beef and pork, and reduced loss of glutathione.[10,11]

Low-temperature irradiation of raw meat did not produce, however, a product which could be stored for considerable periods, because of the continued presence of active proteolytic enzymes. As a result of their activities, undesirable odours and flavours developed within a few weeks.[12] More successful radappertisation of beef, chicken rolls and other meat products was achieved by combining mild blanching to inactivate enzymes with low temperature irradiation in the range of -30 to $-80\,°C$.[9]

Investigation of the nutritional quality showed considerable preservation of thiamine in treated pork, minced meat, frozen pork loin, pork sausage and bacon. The protective effect was greater for thiamine than other vitamins of the B group.[13,14] The vitamin E content of minced pork[15] and the vitamin A content of hog liver[16] were similarly preserved by lowering the temperature during irradiation.

The exclusion of oxygen during irradiation appears to yield conflicting results. Early investigations claimed advantages, although the exclusion of oxygen would affect neither the radiolysis of water nor the reactions of the primary radicals with food constituents nor the direct reactions between the radiation and food constituents. Replacement of air by nitrogen had no significant effect on canned irradiated beef, pork and bologna. Other investigators claimed increased peroxide and carbonyl values in pork, beef and chicken irradiated in oxygen but not in nitrogen atmosphere. Thiamine losses were not affected by oxygen exclusion, while vitamin E was retained much better during storage in a non-oxygen atmosphere.

Fat content, irradiation temperature and storage time are factors likely to influence the results of oxygen exclusion. However, irradiated fats show a tendency for autoxidation in the late post-irradiation period, while peroxide values of irradiated beef, blanched before processing, remained unchanged if vacuum packed in cans or flexible pouches.[17]

Undesirable volatile compounds can be removed by the inclusion of activated carbon in the headspace of containers. Only volatiles present in the headspace are removed, not those present in the interior of the meat.[18] On the whole little value attaches to this practice.

2.4. Effects of Dose-Rate and Post-Irradiation Effects

Dose-rate appears to have little effect on the palatability of irradiated foods, consumer acceptance being similar for roast beef sterilised by γ-rays or accelerated electrons.[9] However, the high dose rates of accelerated electrons promote the recombination of free radicals with consequent less reactions in the treated food and with less destruction of vitamins.[72]

Most chemical changes occur within fractions of microseconds after irradiation but some secondary reactions, such as oxidation by hydrogen peroxide, proceed during the post-irradiation period. Protein solutions showed increasing heat sensitivity, easier denaturation, and the development of more disordered structure, e.g. unfolding of helices, in the post-irradiation period.[4] Autoxidation of fats in the presence of air through the formation of peroxy radicals yields a variety of breakdown products, which in their turn may lead to other reactions, e.g. vitamin E destruction.[19] Protein-rich foodstuffs showed different compositions of volatiles immediately after irradiation compared to 6 months later. Thus irradiated beef lost *n*-alkanes and 1-alkenes on storage and this effect could be reduced by storage in the absence of oxygen.[20] Similar observations were made for vitamin E and vitamin A. Radiation-induced peroxide formation could be largely prevented in the post-irradiation period by adding various antioxidants.[21] The decay of some free radicals in stored irradiated products could be correlated with their water activity and the length of storage by observation of the behaviour of the ESR signals.

2.5. Summary

The chemical analysis of the changes which may occur in irradiated foodstuffs either shows losses in the original constituents or reveals the presence of reaction products. Alterations affecting the structure of the major components have nutritional implications particularly if the levels of essential amino acids and vitamins are lowered. However, changes of the order of less than 10% would be difficult to detect, allowing for normal biological variability.

The formation of reaction products, particularly off-flavours, is of technological importance while the formation of possible harmful compounds is of toxicological interest. The formation of specific compounds permitting the detection of food processing by irradiation would be of regulatory interest. In these areas highly sensitive analytical procedures have been developed permitting the elucidation of some of the reaction mechanisms occurring in foodstuffs as a result of irradiation.

3. RADIOLYTIC EFFECTS IN MEAT AND POULTRY

3.1. Overview

Meat represents the post-mortem state of the flesh of the animal. It consists essentially of muscle cells and fat embedded in connective tissue. Actomyosin, the water-insoluble contractile protein of the myofilaments, constitutes only about 10% of meat. About 6% of meat consists of water-soluble enzymes, present in the fluid around the myofilaments. To preserve meat and retain its flavour processing, such as chilling, freezing, cooking, canning, dehydration, curing or irradiation, is necessary.

The general effects of irradiating meat, as explained earlier, are the production of H˙ and OH˙ radicals in the aqueous phase. These radicals react with the organic molecules present to form secondary free radicals, polymers, peroxides and other radiolytic products. The changes will depend on the absorbed dose. At 50 kGy the myofibrils lose their water-holding capacity, about 13% of the amino acids are lost and a 'wet dog' odour develops, when the irradiated meat is heated. At high doses the triple helix structure of collagen becomes disorganised with the formation of intermolecular crosslinks. Myoglobin is converted to bright red oxymyoglobin, while some green sulphmyoglobin and (mostly) brown metmyoglobin is formed. The lipids develop oxidative rancidity and unsaturated lipids tend to polymerise and also form hydroperoxides and carbonyls. The carbohydrates partially oxidise to gluconic acid and aldehydes. Vitamins C and B are most affected; but no carcinogenic substances have been detected in feeding tests with irradiated meat.[92] The temperature and dose of irradiation also affect the proteases in meat as shown by treating fresh beef, pork and chicken with irradiation doses varying from 30–50 kGy at temperatures ranging from 21 °C to −80 °C. No reduction in beef proteases—but 13% reduction in pork and chicken proteases—has been noted with 20 kGy doses at −80 °C. All three meats showed 90% reduction with doses of 80 kGy at 21 °C and 45% reduction with 37 kGy at −30 °C.[66]

3.2. Changes in Volatiles

High-dose irradiation of meats had been observed to cause off-flavours and off-odours. It was therefore of interest to determine the source of these radiation-induced, odoriferous volatile compounds by studying the effect of irradiation on various fractions obtained by extraction and dialysis. The typical off-odour was found to be associated with the irradiated water-soluble fraction[22] but not with the irradiated, isolated meat fat.[23] Irradiation of lean meats was reported to produce more off-odour than high-fat meat.[24,25] Similarly irradiated pork was more acceptable than veal or beef.[26]

Glutathione and sulphydryl compounds were found to be involved in odour production,[27] particularly as glutathione content decreased on irradiation in beef and pork,[22] while hydrogen sulphide was produced in quantities ranging from 0·1 to 1 mg/100 g beef depending on the temperature and radiation dose employed.[24,28] Ageing of beef itself increased the hydrogen sulphide content, a further rise occurring after irradiation. Similar increases were found in methylmercaptan levels. Levels of both compounds decreased during post-irradiation storage.[29] Further evidence came from irradiating ground beef with added [^{35}S]-glutathione, when 91% of the methylmercaptan and 20% of the hydrogen sulphide contained the ^{35}S label. Similar studies with ^{35}S-labelled methionine showed a transfer of the ^{35}S label to 8·7% of the methylmercaptan and 21·8% of the hydrogen sulphide.[30] Thus most of the sulphur-containing volatiles are derived from the protein part of meat. Irradiation therefore causes a permanent loss of sulphydryl groups in meat. This was measured in pork meat irradiated with doses ranging from 1–25 kGy. The dose response was a double exponential function, because two fractions of sulphydryl groups exist. The radiosensitive fraction represented 20–50% of the total SH content (G value 23) while the radioresistant fraction had a G value of 3–7. The level of sulphydryl groups could therefore be used to indicate radiation treatment.[77]

Volatile amines also contributed to the off-odour. In irradiated beef 92–95% of the total volatile bases were found to be ammonia and most of the rest was methylamine and ethylamine, total amines amounting to about 2·5 mg/100 g at 37 kGy.[31]

Carbonyls were also found to contribute to the off-odour and increases were detected if oxygen was excluded during irradiation. They derived partly from the meat fats and partly from the proteins as demonstrated experimentally for the proteins from beef, pork and chicken. Carbonyls soluble in iso-octane increased in the lipid fraction, particularly in the presence of oxygen, while water-soluble carbonyls were raised in the protein fraction. The carbonyls derived from the proteins were mostly short-chain aldehydes and ketones: those from the lipids were mostly of chain length greater than ten carbon atoms. Irradiation at 19 kGy thus produced about 17 mg iso-octane-soluble carbonyls per 100 g beef, assuming an average molecular weight of 200.[4,32]

The hydrocarbons detected after irradiation originated mainly from lipids by fragmentation of the fatty acid sidechains. Cleavage occurred preferentially near the ester linkage. Some of the hydrocarbons resulted from decarboxylation of the fatty acids and from breaks of the carbon chain. The saturated fats yielded mainly *n*-alkanes and *n*-alkenes from C_1–C_{17}. Terminally unsaturated alkenes and some alkadienes from C_{15}–C_{17}, as well as trace amounts of alkynes, were derived essentially from unsaturated fats and oleic acid. Some hydrocarbons also originated from the sidechains of amino acids. Reactions of free radicals with oxygen led to the production of hydroperoxides and then of carbonyls.[38,68] Hydrocarbon yields from pork meat increased linearly with dose irrespective of the presence of oxygen, amounting to about 10 mg/100 g meat.[37] Interestingly, heated lard produced C_4–C_7 cyclohexenes while irradiated lard yielded specifically C_{14}–C_{17} dienes. Heating produced more total hydrocarbons than irradiation with 30 kGy.[42] However, cooked pork fat, irradiated under vacuum with 10–50 kGy, lost preferentially conjugated dicarbonyls.[86]

Thus highly sensitive analytical procedures have identified a very large number of volatile compounds in irradiated beef, veal, mutton, lamb, pork and chicken. The qualitative and quantitative distribution depended largely on temperature and radiation dose, total volatiles increasing almost linearly with irradiation dose.[33,34,35,68] Table 1 lists some of the volatiles and their abundances.

Larger quantities of volatiles were identified from isolated irradiated beef and pork fat than from fat in meat, showing the loss of protection by meat proteins.[37]

Investigation of the various fractions of meat has therefore revealed that the lipids yielded mainly *n*-alkanes and *n*-alkenes, the protein fraction gave rise to various S-compounds and some alkylbenzenes, while the lipoprotein fraction produced aliphatic hydrocarbons and S-compounds carrying the characteristic irradiation odour.[38,39,40,41]

3.3. Interaction of Meat Components on Irradiation

The reaction products of proteins and lipids in irradiated meats are generally the same as are found in the irradiated isolated components. However, some interactions must occur because the special irradiation odour develops only in the lipoprotein fraction. So far only butyramide has been identified as a specific radiolytic interaction product.[5] Some interaction products with antioxidant properties also appear to be formed, as evidenced by the good oxidative stability of irradiated meats and by low

TABLE 1
COMPOSITION AND ABUNDANCE OF VOLATILES IN IRRADIATED BEEF

Compounds found among volatiles				*Relative abundance in ppm*			
alkanes	C_1–C_{20}	alkanols	C_2–C_6	alkanes	12	alcohols	1·0
alkenes	C_2–C_{20}	alkanals	C_2–C_6	alkenes	14	alkanones	< 1·0
alkynes	C_2–C_{20}	alkanones	C_2–C_6	alkanals	1·5	alkylbenzene	< 1·0
alkadienes	C_4–C_{20}	esters	C_3–C_{18}	S-compounds	10	esters	< 1·0
thioalkanes	C_3–C_6	thiols	C_1–C_4				

Modified from Merritt *et al.*[68]

levels of thiobarbituric acid-reactive substances,[43] especially if fat content was high. Oxygen uptake was also reduced after irradiation of beef, pork and chicken[44] particularly if stored in the absence of oxygen. Heat-sterilised beef also develops antioxidants, probably associated with the aminohexose reductones formed on heating, an effect also seen in irradiated and heated amino acid–sugar solutions. Antioxidant action is noted particularly in ground meat but is less obvious in whole meat; the latter should be stored under exclusion of oxygen.[4]

3.4. Changes in Protein Structure

Closer investigation of the protein aggregates formed by irradiation indicated that the cross-links were covalent in nature and that hybridisation of proteins could occur, without affecting any enzymatic activity possessed by the proteins involved.[46] Proteins in aqueous solution aggregated proportionally to the protein concentration and the irradiation dose. Aggregation was reduced by the addition of carbohydrates; but oil induced aggregation even at 10 kGy doses. Emulsified oil had no effect on aggregation.[69] Sarcoplasmatic proteins of pork and beef aggregated to a molecular weight of about 465 000 by irradiation, with consequent reduced solubility.[49]

The reduced water-binding capacity of muscle protein following irradiation leads to increased drip loss in meats, although even in irradiated meat the water-binding capacity rises as the pH increases.[45] Heat denaturation has effects similar to irradiation,[47] e.g. splitting of hydrogen bonds to liberate basic groups and decarboxylation of free amino acids. Freeze drying has also many parallel effects; however, there are differences between the effects of these three processes.[114] Discrepancies have also been reported regarding the effects of irradiation on the tenderness of meat, but these may well be related to differences in experimental conditions and the period of ageing before irradiation is applied.[48]

Haemoglobin and myoglobin contribute to the colour of meat and respond similarly to irradiation. When fresh meat is exposed to air, oxymyoglobin is formed and this eventually converts to metmyoglobin. Cooking produces denatured metmyoglobin while curing results in nitrosomyoglobin. High dose irradiation in the absence of air produces a bright red colour in fresh meat, cooked meat and cured meat due to the reduction of globin-myohemichromogen. Subsequent oxidation in air forms brown globin-myohemichromogen.[104] These pigments are closely related though not identical with oxymyoglobin, changes probably having occurred in the globin portion due to aggregation and helix unfolding.[50]

Conflicting results have been reported regarding various enzyme activities in beef muscle and lamb's liver following irradiation at various dose levels. Isoenzymes showed unequal radiation resistance. Lysosomes release, differentially, various enzymes on irradiation without being disrupted.[52] Irradiation may have opposite effects on tissue enzyme activities, low doses often increasing activity by liberation of enzymes from lysosomes, while high doses inactivate enzymes.[53] Similar effects were noted with mitochondrial enzymes.

Irradiation of proteins initiates reactions leading to the formation of ionic and free radical intermediates, which react further to yield, eventually, stable products. The reactions occurring include deamination, decarboxylation, conversion of -S–S- linkages to -SH, oxidation of -SH groups, valency changes of metal ions, cleavage of peptide bonds and aggregation. These events have been followed by pulse radiolysis and ESR spectroscopy.

The aromatic and sulphur-containing amino acids are most susceptible to irradiation. However, the reactions of single amino acids and of amino acid mixtures differ in their reaction rates and in the end products. Irradiation of amino acid mixtures induces the formation of α, α'-diamino acids.[70]

Phenylalanine reacts with OH radicals to yield *o*-, *m*- and *p*-tyrosine, of which only *p*-tyrosine occurs naturally. The presence of *o*- and *m*-tyrosine could thus be regarded as a marker for irradiation. Irradiation also acts directly on phenylalanine, causing C–C bond scission with the formation of benzene and toluene.[72]

Peptides undergo primary deamination by electron transfer or secondary deamination with bond scission. Aromatic sidegroups are protonated, while ammonia, fatty acids, keto acids and amides are also produced.[70]

The OH radical reacts with the C–H bond of optically active aliphatic peptides to form α-peptide radicals, which then dimerise and racemise. This radiation-induced racemisation could lead to the formation of D-acids, some of which are known to be toxic. However the amounts of D-acid formed would be very small and unlikely to be a hazard. These changes could however contribute to loss of digestibility.[72]

Deeply buried groups in proteins are less accessible to irradiation. The conformation of proteins, whether random coils or helix, also influences the reaction rates. Radical sites within the protein tend to migrate, forming aromatic and sulphur groups as intermediates. C-radicals, formed in the main chain at sub-freezing temperatures, tend to decay on thawing. In some proteins the molecular weight decreases as a result of peptide bond scission, thereby increasing the solubility and decreasing the tensile strength of the protein fibres. In fibrillar proteins intermolecular cross-links form which increase the molecular weight, decrease the solubility and increase the tensile strength. Globular proteins unfold and aggregate, while fibrous proteins degrade, thereby becoming denatured without proteolysis.[70] The behaviour of collagen after irradiation may vary with the species. However, oxygen and free radical scavengers inhibit cross-linking in collagen, while high concentrations favour cross-linking as seen in myoglobin.[6,45]

Treatment of lean pork with doses of 2, 10 and 50 kGy affected the amino-transferases (AT) of muscle proteins. At 2 and 10 kGy the activity of GOT (aspartate AT) and GPT (alanine AT) increased in tissue and sarcoplasm, i.e. the muscle juice. At 50 kGy the activity of both ATs was reduced. Irradiation at 50 kGy also damages the cell membrane partially releasing peptide fragments from myosin but not from actin.[103]

3.5. Changes in Other Meat Constituents

Conflicting results have been reported for creatinine, creatine and non-protein nitrogen in irradiated beef. Carnosine and histidine appear to form compounds on irradiation which give colour reactions with ninhydrin.[36] Glycogen is reduced by irradiation of fresh meat just as by roasting, while fatty acids and phospholipids remain unaffected. Few investigations have been made on nucleic acids in meat. In chicken total RNA increased at the expense of DNA following irradiation.[51]

3.6. Summary

The direct extrapolation of the radiation effects on proteins to meats requires care because of the influence of other food components on these reactions and of possible interactions between the radiolytic products of all meat components. Chemical effects, effects on nutrients and on amino acid profiles are negligible below a dose of 10 kGy. Radiolytic mechanisms have been demonstrated convincingly to be common for various meats, and the effect of processing parameters on these products is largely predictable. Low-temperature irradiation of meat leads to minor but still acceptable changes. Free radicals are absent, meat proteins show little structural alterations, and no significant loss of amino acids occurs. The volatile radiolytic products from various meats depend on meat composition and are largely predictable as to nature and yield. Hence evaluation at one dose can be used when considering other doses.

Studies on beef and pork meat after irradiation have shown the same changes as are seen with mild heat treatment, e.g. decreased solubility of sarcoplasmic protein, a decrease in myosin and increased stability of actin. The total sulphydryl content did not change: only in the sarcoplasmic fraction did it drop significantly at high irradiation dose or if irradiated in the presence of oxygen.

4. NUTRITIONAL ASPECTS OF MEAT IRRADIATION

A survey of the nutritional aspects of irradiated meat and protein showed that the nutritional effects of ionising radiation treatment, as envisaged commercially, did not differ from those of other preservation methods, especially if treatment was carried out at low temperatures in the absence of oxygen. Animal feeds sterilised with doses of 50–60 kGy retained the biological value of the protein as shown by the satisfactory maintenance of germfree rat and mouse colonies over years on irradiated feed. The metabolisable energy value of the rodent diets remained unaffected by the irradiation treatment. Irradiated pork (28 kGy), stored subsequently for 1 year at room temperature, had the same digestibility as unirradiated pork when eaten by volunteers. The overall digestibility of lard also remained unaffected.

Radappertised beef (47–71 kGy), pretreated to inactivate enzymes and irradiated at −40 °C, was compared to heat sterilised and frozen, enzyme-inactivated beef. No significant destruction of cystine, methionine or tryptophan was noted immediately after irradiation and on re-examination after 15 months storage. Chicken meat, irradiated with 6 kGy, stored for 6 days at 5 °C and then cooked, showed no difference in lysine availability or protein efficiency ratio when compared to untreated poultry meat. Ground pork irradiated with 58 kGy showed less than 10% destruction of pantothenic acid and no destruction of folacin.[78]

Beef treated by freezing, thermal processing. γ-irradiation or electron-irradiation had no antithiamine properties when fed to thiamine-deficient rats. Growth, erythrocyte transketolase activity and the amine pyrophosphate effect were similar for all meat-fed groups.[54] Irradiation sterilised chicken behaved similarly.[55] Irradiation of meat at −40 °C does not cause significant losses of amino acids and little structural alteration of proteins.[114]

The biological value of proteins was found to be little affected even by high doses of irradiation as demonstrated by the adequate growth of chickens fed on irradiated material.[87] However, model experiments with irradiated fat–casein mixtures produced some evidence of reduction in net protein utilisation. Mixtures of casein, corn oil and coconut fat in the ratio 11:2:2 were either heat sterilised or irradiated at 50 kGy and stored for 12 weeks under normal atmospheric conditions. Feeding of these mixtures to rats at a 10% protein level reduced net protein utilisation. No effects on the nutritive value of casein were noted if irradiated casein alone was fed or irradiated casein with glucose or irradiated casein with starch. No significant effect was also seen if the irradiated casein–fat mixture was fed within 2 weeks of irradiation or if the lipid was extracted after 12 weeks storage and replaced by untreated fat. Irradiation of the complete diet, followed by 12 weeks storage, also reduced net protein utilisation. The oxidation of lipids by irradiation probably caused the formation of carbonyls. These breakdown products reacted with the protein to decrease the available lysine and damage other amino acids. Irradiation also forms high molecular weight protein aggregates particularly on storage. Both an increase in carbonyls and of protein aggregates could be demonstrated by analysis.[88]

When a casein–fat mixture in the ratio 11:4 was irradiated with 50 kGy in the absence of oxygen, stored for 12 weeks and then fed to rats no reduction in net protein utilisation was found. Similarly using a saturated fat, irradiated and stored under aerobic conditions, no effect on net protein utilisation was noted. If, however, a sunflower oil was used, a drastic reduction in net protein utilisation and growth inhibition was found. These effects could not be reversed by vitamin E. Carbonyls were grossly elevated and available lysine was also reduced when sunflower oil was substituted in the casein–fat mixture.[89]

The effect of irradiation at 50 kGy or of heat sterilisation at 100 °C for 60 min on casein was examined in growing rats by means of nitrogen balance studies. Irradiation of casein alone, of casein with added glucose, and of casein with added starch had no effect on digestibility or net protein utilisation. Heat sterilisation significantly reduced both digestibility and net protein utilisation in the presence of glucose. Available lysine was also reduced.[90]

The biological value of irradiated protein–lard mixtures was compared with that of irradiated protein and irradiated lard. The irradiation dose was 40 kGy, the materials were vacuum packed and subsequently stored for 6 to 12 weeks at room temperature. Feeding to rats showed no effect of irradiation or storage on the biological value of the protein. Similar experiments were carried out using irradiated and non-irradiated actomyosin, isolated from pork and denatured by heat, again in mixtures with irradiated and non-irradiated lard. Rats were fed with mixtures immediately after irradiation, or with mixtures stored for either 6 or 12 weeks. No effect on net protein utilisation was noted.[91] In another experiment, in which casein, freshly irradiated with 20 kGy, was fed to rats, no depression of nitrogen turnover was noted but growth was reduced by 40%. The effect disappeared after 35 days storage.[67] Irradiation of pork with doses of 45–56 kGy at −30 °C had a less destructive effect on thiamine, riboflavin, niacin and pyridoxine than heat treatment.

Although carbohydrates are depolymerised by hydrolysis and are oxidatively degraded their biological availability remained unaffected.[159] Similarly, although lipids are oxidised, degraded and decarboxylated, especially if they contain unsaturated fatty acids, no effect on the digestibility for man of pork fat treated with 28 kGy and stored for 1 year was detected.[141] However, lard irradiated at 56 kGy, was more slowly absorbed by dogs than untreated material.[142]

Proteins and amino acids are only minimally affected by doses up to 70 kGy. Irradiated meat fed to rats was of similar digestibility, biological value, net protein utilisation and amino acid composition as untreated meat.[143] Investigation of the effect of irradiation at room temperature on soluble proteins and LDH isoenzyme activities showed that soluble proteins decreased at all doses and proteins aggregated at 50 kGy. Doses of 25 kGy reduced the activity of all isoenzymes.[120]

Thiamine retention in radappertised ham increased with lower doses and lower temperatures during irradiation. Similar effects were seen in pork loin when compared to thermal processing. Thiamine, riboflavin and pyridoxine are reduced while niacin is not affected. Vitamin B_{12}, *p*-aminobenzoic acid, pantothenic acid and folacin are more easily destroyed in aqueous solution than in food.[144] Pantothenic acid and folacin are reasonably resistant.[145] Carotene is radiosensitive but can be protected by ascorbic acid while 60–70% of vitamin A and of the tocopherols are destroyed by 4 kGy.[112] Vitamin D was also sensitive to irradiation; but not vitamins K_1, K_3 or K_5.

5. MICROBIOLOGICAL ASPECTS OF MEAT IRRADIATION

Fresh red meat is sometimes contaminated with pathogens through commercial processing and in distribution. Most of the microflora has been found to be radiation sensitive, so that central processing by irradiation would achieve a longer shelflife. The residual flora after irradiation has no unique bacteriological significance. Irradiation with 2 kGy destroyed most known pathogens and allowed storage for 8 days at 2 °C without quality change. Irradiation with 0·68 kGy destroyed 99·9 % of the total microflora of ground beef and pork sausage but was insufficient to remove pathogens.[73]

Moraxella-Acinetobacter, a gram-negative coccobacillus, is known to occur on various types of unprocessed beef and poultry. Certain strains are radioresistant, some are also sufficiently heat resistant to survive pre-irradiation heat treatment for enzyme inactivation.[56] The latter strains are also haemolytic, which is a common marker for pathogenicity.[57,58] However, experiments with red meat and poultry showed that these bacteria were neither a health hazard nor that they caused organoleptic changes.[59] The application of heat, e.g. 65–75 °C, injured rather than inactivated radiation-resistant cultures of *Moraxella-Acinetobacter* in minced beef. Thermally-injured cells were then more sensitive to subsequent irradiation at −30 °C. Thus, although *Moraxella-Acinetobacter* cells are more resistant to irradiation than *Clostridium botulinum* spores, earlier thermal processing redresses the balance and contributes to the elimination of these micro-organisms.[74] Radiation-resistant *Moraxella-Acinetobacter* will only grow in ground beef or chicken if the a_w (water activity) is at least about 0·99.

Radurisation of fresh meat with doses of 1–2 kGy improves the microbial contamination only to a small extent. Gram-positive anaerobic organisms tend to predominate in the surviving flora. Gram-positive, catalase-negative cocci are largely eliminated but gram-positive, catalase-positive rods were found in all irradiated meats. New health hazards therefore do not arise as a result of radurisation.

Radicidation of fresh red meat and poultry with doses of 3–5 kGy aims at eliminating the pathogens carried by these foods. The health problems created by food-borne diseases are being increased by the continued expansion of mass breeding and fattening of animals, by the mass production and increasing processing of foods, the encouragement of tourism and the expanding trade in food and animal feeds.

Salmonella is the principal pathogen carried by poultry, ground red meat, hamburgers and fresh pork sausages. Salmonellosis is thus a cause of major morbidity, particularly through the secondary contamination of kitchen utensils, work surfaces and hands. *Campylobacter jejuni* is recognised today as another important cause of gastro-intestinal infection. It is carried essentially by poultry, but also by meat and milk. *Cl. perfringens* has also been recovered from meat and meat products, especially turkeys, large red meat portions and meat pies (if rewarmed).

Measures other than irradiation may be used to control food-borne disease. For example, raising food animals free from pathogens can be done on a small scale, although recontamination at slaughter may nullify the effort. Surface decontamination of raw meat with steam or hot water or canning are other alternatives. Irradiation up to doses of 5 kGy has been demonstrated to reduce the viable non-sporing pathogens and has the advantage of permitting the treatment of packaged food and of low energy expenditure. Viruses are not affected by these doses of irradiation.

Radappertisation to commercially sterilise enzyme-inactivated packaged beef must achieve a 12 D reduction in the number of viable spores of *Cl. botulinum*, i.e. from 10^{12} to 10^{0} spores. D is defined as the dose required to reduce the population of bacterial spores to 90%.[136] Inoculated-pack studies using the most resistant spores of *Cl. botulinum* strains 33A, 36A, 40B, 41B and 53B[137] are carried out to determine D for each individual food. Examples of 12 D doses are 41 kGy at −40 °C for beef, 44 kGy for pork, 43 kGy for chickens and 15 kGy for bacon cured with 40 mg/kg sodium nitrite.[110]

Inoculation experiments have shown that *Campylobacter jejuni* can be eliminated by irradiation of ground beef or ground turkey with 2 kGy. Salmonella infection carried by poultry is a mortality hazard particularly to high-risk individuals in hospitals. This hazard can be eliminated by irradiation.[62]

Viruses are generally radiation-resistant under experimental conditions.[60] However, irradiation might possibly reduce the number of infective virus particles. Foods processed by heat treatment above 55 °C before or after irradiation probably present a much reduced risk compared to raw foods of animal origin. Irradiation is thus of comparatively little value as a preventive measure against dangerous animal virus diseases.[61]

6. TECHNOLOGICAL ASPECTS OF MEAT IRRADIATION

6.1. Radurisation

The radurisation of meats and poultry by low dose irradiation aims at delaying the onset of microbial spoilage, thereby extending the product life through reduction of the initial microbial population. For reasons of microbiological safety and maintenance of quality it must be combined with refrigeration. The microbiological improvement corresponds to a reduction by only one log cycle (1 *D*). The technology is useful principally for the fresh red meats of beef, veal, pork and lamb as well as fresh chicken. Processed perishable meats, e.g sliced cold cuts, fresh, cooked, smoked and cured sausages, are also suitable for radurisation. However, the conventional vacuum packaging of perishable processed meats under exclusion of oxygen equally delays microbial spoilage for 30 days or more, a period adequate for distribution and marketing. Radurisation of processed meats is thus rarely justified.[105]

The storage life of fresh meats and poultry is about 3 days despite good refrigeration, making it difficult to centralise the distribution of prepackaged items. Hence non-irradiated meats are presently transported in bulk, the final cutting and packaging being done at the retail sale point. Vacuum packaging of wholesale pieces is now feasible but the colour of frozen retail cuts in vacuum packs is unacceptably dark.

Radurisation has the disadvantage of not delaying the oxidation of myoglobin to metmyoglobin; it does not reduce drip from the cut surface, and it does not prevent lipid oxidation, so that off-flavours can develop. To avoid these changes during the prolonged storage for up to 3 weeks of radurised meats requires the addition of up to 0·5% sodium tripolyphosphate against drip and the use of oxygen-permeable, moisture-impermeable plastic wrapping. The irradiation dosage required is 1–2 kGy at 0 °C to 10 °C. Bulk packages and retail packages must be stored at 0–5 °C.[129]

Recent experience has shown that doses of 0·25–1 kGy suffice for aerobic storage, while 1–4 kGy doses are useful for anaerobic storage. Doses of 4 kGy are, however, unsuitable for mutton as they induce off-flavours.[130]

Higher doses are difficult to apply to fresh meats at ambient temperature as off-flavours arise at 1·75 kGy with pork, at 2·5 kGy with beef and chicken, and at 6·25 kGy with lamb.[106] The use of sub-freezing temperatures during irradiation with subsequent thawing is not practicable for these products. Irradiation doses are also restricted by possible colour changes. Doses of 1 kGy do not induce lipid oxidation. Radurisation allows therefore 24 days for preparing and distributing meat for retail sale.

Irradiation with doses of 3–6 kGy can prolong the storage life of refrigerated meat from 10–15 days to 30–50 days by decreasing the bacterial population load, but the induced aggregation of the myofibrillar protein makes the protein less extractable and lowers the quality of the meat. Adrenalisation as pretreatment, as well as the addition of salt, sodium tripolyphosphate and ascorbic acid before irradiation, has been claimed to stabilise water retention, especially in pork.[65]

6.2. Radicidation

This process involves larger doses of irradiation but can also be applied to raw meat and poultry. It is designed to remove pathogenic organisms (e.g. Campylobacter and Salmonella) from these products, which are the common carrier foods. Irradiation aims at a 7*D* (7 log cycle) reduction in the non-spore forming microbial load which would eliminate Salmonella and other pathogens.

If poultry is kept at 0–5 °C the shelflife of untreated birds is 2–4 days and some 60 % are known to carry Campylobacter or Salmonella. Irradiation with doses of 3 kGy will extend shelflife to a maximum of 15 days, and 5 kGy will give a maximum of 20 days shelflife without any off-flavour. About 4·5 kGy will give a salmonella-free product, while chilled and deep frozen poultry treated at these doses will also be free of Campylobacter.[155]

Minced red meat usually carries pseudomonas; gram-negative enterobacteria, lactic acid producers', and about 17 % is also infected with Salmonella. Untreated meat, kept at 4 °C, could be stored for at most 3 days. Irradiation with 3 kGy reduces Salmonella by at least 5 log cycles and eliminates pseudomonas, enterobacteria and lactic acid formers as well as any staphylococci giving a possible shelflife up to 2 weeks. However, the need for freezing the meat before irradiation and subsequent thawing makes it an expensive treatment.

The keeping period of offal could be doubled by irradiation with 6 kGy.

Mechanically separated poultry meat can be treated with 2·5 kGy if packed in polythene bags and kept at −18 °C in order to achieve a storage life of 12 days at a storage temperature of 0 °C.[156] Prepacked ground beef, minced meat, hamburgers and fresh pork sausages may need from 2–7 kGy for the elimination of pathogens.[155] The loss of hedonic properties such as flavour, aroma and colour can be minimised by chilling the food at the time of irradiation. Inactivation of helminths requires doses of the order of 0·2–0·6 up to 1 kGy. *Trichinella spiralis* is probably the least resistant, requiring doses of 0·3–1 kGy irrespective of the muscle site or age of cyst, while *Cysticercus bovis* and *Taenia solium* respond to the average dose of 0·4–0·6 kGy. Lower doses, of the order of 0·3 kGy, together with storage at 2 °C for 7 days have also been reported to be fatal to *Cysticercus bovis*. The procedure may be important for countries with helminth infections although deep freezing will also kill the parasites. To avoid off-flavour subfreezing temperatures would be needed during irradiation.[64] Doses of 0·02–0·03 kGy can also inactivate *Trichinella spiralis* but do not prevent the initial intestinal phase and are therefore not an adequate health measure.[63]

The role of foods in transmitting viral diseases is not clear. Raw meat is, however, a carrier for the virus of foot-and-mouth-disease. Being a very small virus it requires 4·8–6·1 kGy for inactivation at the 1 *D* log cycle reduction level. At least 20 kGy would be required for treating infected meat.

6.3. Radappertisation

Radappertisation is the alternative to commercial preservation by thermal sterilisation, freezing and drying. Thermal processing of non-acid foods suffers from loss of juices, requires small containers because of poor heat conductivity, and results in unsatisfactory texture and changes in flavour. Deep-freezing is expensive in energy, frozen food requires thawing before use, and a power supply is essential. Drying and freeze drying can be applied only to powdered or thinly sliced products, and involves high energy costs.

The advantages of radappertisation are that the treated products closely resemble fresh food, that there is minimal loss of juice, that there are no limitations on the size of the containers, and that sterility is assured. Radappertisation of meat and poultry requires exposure in sealed containers and doses of 25–70 kGy at −30 to −40 °C to kill all spoilage organisms and achieve commercial sterility. With canning it yields products which may be stored for years at ambient temperature and which are still acceptable, with minimal changes in texture, flavour, odour and colour. Irradiation must be combined with heating meat to an internal temperature of 65–75 °C to inactivate the proteolytic enzymes which would spoil meats during non-refrigeration storage.

The method has prospective value for prepackaged cuts of beef, lamb, pork and poultry, corned beef, sliced ham, pork sausage, bacon, luncheon meats, chicken and turkey slices and shrimp. These products have been consumed by astronauts and in submarines in the form of meat sandwiches, ham and bread, corned beef and turkey slice sandwiches, and as beefsteak. Similar meals, including precooked hamburger patties, pork chops, pork sausages and chicken breast, have been supplied to patients in germ-free isolation wards. Beef products still need improved flavour and taste. There is the added attraction of being able to store treated food in supermarkets on the open shelf with savings in freezer energy. The flexibility in the packaging size allows prepackaging at variable serving size, thus offering advantages to campers, truck drivers and astronauts as well as to institutional feeding, fast food chains, airline catering, and for stockpiling emergency food supplies.

The occurrence of off-flavours and off-odours, of adverse colour changes, changes in texture, and in the nutritional value after irradiation necessitates the use of vacuum sealing against oxygen attack and irradiation at low temperatures of the order of −20 to −40 °C. The process is not suitable for food consumed raw or destroyed by freezing and thawing e.g. fresh fruit, but can be applied to meats, poultry and meat products. These foods remain palatable, wholesome and nutritious after storage for years without refrigeration. Some stored foods, e.g. roasts and luncheon meats, cannot be produced by any other preservation process. Radappertisation also allows the avoidance of chemical preservatives, e.g. ethylene oxide, and a reduction in the use of sodium nitrite, thereby lowering the formation of toxic nitrosamines in cured meats.[71] Nitrite is needed for the characteristic flavour and colour of bacon, ham and other cured meats. It also protects these products against the development of *Cl. botulinum* spores. Irradiation in combination with 550 mg/kg ascorbate would permit the reduction of nitrite in bacon from 120 mg/kg to 20 mg/kg or less. Such products have been shown to contain after frying no residual nitrite and about 1–3·6 μg/kg nitrosopyrrolidine.[139] Shelf-stable cured ham can be produced despite the reduction of nitrite from the normal 156 mg/kg to 25 mg/kg. Similarly nitrate can be lowered from the usual 500–700 mg/kg to 100 mg/kg, while the addition of 275 mg/kg ascorbate or erythorbate ensures a stable colour, flavour, odour and taste after irradiation with 35–37 kGy.[97] Nitrite can also be eliminated from radappertised corned beef and frankfurters but the flavour and colour are slightly altered compared to the nitrite-cured products.[97,151]

The new technology requires inactivation of autolytic enzymes in meats. The use of a maximum tolerable dose of 70 kGy destroys between 25–50 % of the enzyme activity without serious loss of palatability and loss of texture. The application of a reliable method for enzyme inactivation is therefore essential. Only heat treatment has been found to be satisfactory; chemicals were unreliable.[13] The efficiency of enzyme inhibition has been assessed by measuring the radioactivity liberated from $K^{14}CNO$-labelled haemoglobin by the residual meat proteases.

The loss of natural juices from ground meat, which normally reaches about 35 % after enzyme inactivation, can be minimised to 10–15 % by the addition of 0·75 % sodium chloride and 0·25–0·5 % sodium tripolyphosphate. The slightly drier final product, when reheated for serving, has been found to be organoleptically acceptable.[135] The exclusion of atmospheric oxygen by using vacuum sealed packaging also provides a means to exclude moisture and contaminating micro-organisms. This is usually achieved by evacuation. Irradiation is then carried out at −40 °C (for gamma ray sources) or at −20 °C (for electron sources). Under these conditions the organoleptic and nutritional qualities and the digestibility of the food remain unaffected, although the resistance of any contaminating *Cl. botulinum* is somewhat increased. However, the heat treatment for enzyme inactivation also inactivates most of the food-borne viruses and those vegetative bacteria, e.g. *Moraxella-Acinetobacter*, which are more resistant to radiation at low temperatures. Thus the final product is free of public health hazards and similar to thermally canned foods.

Inoculated pack studies, using between 10^6–10^7 spores per can or pack, and irradiation at −30 °C with variable doses, showed that the minimum sterilising dose was about 25 kGy for beef, while the 12 *D* dose varied between 37 and 46 kGy.[110] The 12 *D* dose for cured ham was about 35 kGy but at lower doses viable *Cl. botulinum* spores could survive without, however, producing any toxin.[96] For low-nitrite ham, treated with 25 mg/kg nitrite and 100 mg/kg nitrate, the 12 *D* dose was 32–38 kGy at

TABLE 2
MINIMUM RADIATION DOSE FOR STERILISATION[d]

Food	*Irradiation temperature (°C)*	*ESD*[b] *(Mrad)*	*MRD (Mrad)*[a] Schmidt–Nank[c] *Range*	*Mean*
Bacon	5–25	1·5–2·0	1·94–2·65	2·30
Beef	−30 ± 10	3·5–4·0	4·44–4·99	4·66
Beef	−80 ± 10	4·0–4·5	5·20–6·13	5·70
Chicken	−30 ± 10	3·0–3·5	4·34–4·62	4·48
Codfish cake	−30 ± 10	2·5–3·0	3·14–3·19	3·17
Corned beef	−30 ± 10	2·0–2·5	2·36–2·57	2·47
Ham	5–25	3·0–3·5	1·79–3·20	2·90
Ham	−30 ± 10	3·0–3·5	2·87–4·37	3·66
Pork	5–25	3·5–4·0	4·19–4·99	4·56
Pork	−30 ± 10	3·5–4·0	4·57–5·68	5·09
Pork sausage	−30 ± 10	1·5–2·0	2·40	2·40
Shrimp	−30 ± 10	3·5–4·0	1·99–5·12	3·72

[a] Minimum radiation dose in megarads, $D \times 12$.
[b] Experimental sterilising dose, range in megarads.
[c] $D = \text{Mrad}/(\log M - \log S)$.
[d] Compiled by Mr Abe Anellis, US Army Natick Laboratories.

−30 °C. No growth was found after post-irradiation incubation for 6 months at 30 °C.[98] Some minimum sterilisation doses are listed in Table 2.

The reduction in the 12 *D* dose by the addition of sodium chloride and sodium tripolyphosphate is balanced by the increased dose requirement of using −40 °C as irradiation temperature. Packaging in appropriate materials contributes to the safety of irradiated foods. Tinplate can resist doses up to 75 kGy at −90 °C.[138] Plastic flexible films and laminated pouches with aluminium foil as middle layers are good barriers for moisture, light and oxygen, for doses up to 60 kGy. Table 3 lists the food packaging materials approved by the FDA of the US for irradiated foods.

The organoleptic quality of radappertised meats was reported to be acceptable. Early investigations had shown that carcass grade, temperature at the time of irradiation, and irradiation dose affected the flavour, tenderness and the development of mercaptans, of soluble collagen and extractable non-protein nitrogen in beef-steaks stored for 2 years.[111] Later studies, using a temperature of −30 °C and irradiation doses of 45–50 kGy established this technology as acceptable for the consumer.[121] It is important to emphasise that the starting materials have to be of good quality and sanitary condition because irradiation can only arrest spoilage but not change quality for the better.

In general it is necessary to remove inedible material, to debone and to trim off the excess fat. After inactivating proteases by heating to internal temperatures of 65–80 °C, the hot foods are usually vacuum packed, frozen at once to −40 °C and irradiated at temperatures between −20 and −40 °C. If pretreatment with 0·25–0·5 % sodium tripolyphosphate and 0·5–1·5 % sodium chloride is employed, then meats should be held about 2 h at 1 °C–4 °C, followed by cooking for 12 h at up to 70 °C. This procedure is usually applied to meat chunks which are subsequently pressed into shape by appropriate casing before the heat treatment. The partially solubilised muscle protein forms a skin of denatured protein on the surface, reducing the loss of juice to a maximum of 20 %. Subsequently the casing is replaced by the final packaging before irradiation.

Combined treatment of meat with heat and ionising radiation has been suggested as an alternative to full radappertisation. Thus raw meat, heated to 77 °C for 20 min followed by irradiation with 4–6 kGy will store satisfactorily at 37 °C without spoilage. Semi-finished meat products, immersed for 5 min in vegetable fat at 130–160 °C and followed by 6 kGy irradiation, can be stored for 30 days at room temperature while retaining good organoleptic properties. The vacuum-packed product can be kept at 4 °C for up to 8 months. Infrared grilling until cooked, followed by

TABLE 3

APPROVALS BY THE US FOOD AND DRUG ADMINISTRATION FOR RADIATION PRESERVATION OF FOOD PACKAGING MATERIALS[a]

Packaging material	*Petitioner*	*Source*	*Dose*	*Food additive petition no.*	*Filing*			*Regulation*		
						Fed. Reg.			*Fed. Reg.*	
					Date	*Vol.*	*Page*	*Date*	*Vol.*	*Page*
Nitrocellulose coated cellophane	AEC	gamma	1 Mrad	1297	2-8-64	29	2 318	8-14-64	29	11 651
Glassine paper	AEC	gamma	1 Mrad	1297	2-8-64	29	2 318	8-14-64	29	11 651
Wax coated paperboard	AEC	gamma	1 Mrad	1297	2-8-64	29	2 318	8-14-64	29	11 651
Polypropylene film with or without adjuvants	AEC	gamma	1 Mrad	1297	2-8-64	29	2 318	8-14-64	29	11 651
Ethylene-alkene-l copolymer	AEC	gamma	1 Mrad	1297	2-8-64	29	2 318	8-14-64	29	11 651
Polyethylene film	AEC	gamma	1 Mrad	1297	2-8-64	29	2 318	8-14-64	29	11 651
Polystyrene film with or without adjuvant substances	AEC	gamma	1 Mrad	1297	2-8-64	29	2 318	8-14-64	29	11 651
Rubber hydrochloride with or without adjuvant substances	AEC	gamma	1 Mrad	1297	2-8-64	29	2 318	8-14-64	29	11 651
Vinylidene chloride vinyl-chloride copolymer film (Saran wrap)	AEC	gamma	1 Mrad	1297	2-8-64	29	2 318	8-14-64	29	11 651
Polyolefin film with or without adjuvant substances or vinylidene chloride coatings (Saran)	AEC	gamma	1 Mrad	5M1674	7-30-65	30	9 551	3-19-68	33	4 659

Polyethylene terephthalate with or without adjuvant substances or vinylidene chloride coatings (Saran) or polyethylene coatings	AEC	gamma	1 Mrad	5M1674	7-30-65	30	9 551	3-19-68	33	4 659
Nylon 11	AEC	gamma	1 Mrad	6M1820	9-8-65	30	11 400	3-19-68	33	4 659
Vinylidene chloride copolymer (Saran) coated cellophane	AEC	gamma	1 Mrad	5B1670	2-18-65	30	9 116	6-11-65	30	7 599
Vegetable parchment	US Army	gamma	6 Mrad	5M1622	1-15-65	30	547	3-12-65	30	3 354
Polyethylene film with or without adjuvants	US Army	gamma	6 Mrad	5M1645	7-21-65	30	9 116	6-10-67	32	8 360
Polyethylene terephthalate with or without adjuvants	US Army	gamma	6 Mrad	5M1645	7-21-65	30	9 116	6-10-67	32	8 360
Nylon 6 films with or without adjuvants	US Army	gamma	6 Mrad	5M1645	7-21-65	30	9 116	6-10-67	32	8 360
Vinyl chloride–vinyl acetate copolymer film with or without adjuvant	US Army	gamma	6 Mrad	5M1645	7-21-65	30	9 116	6-10-67	32	8 360
Kraft paper	US Army	gamma	50 000 rad	7M2172	5-23-67	32	7 877	7-19-67	32	10 567

[a] Compiled by Mr Frank Leone, US Atomic Energy Commission.

irradiation at 0·8 kGy and vacuum packing yields a product which can be stored at 4–5 °C for 6 months.[65] Boiling or grilling combined with irradiation at 1·5–3 kGy permitted storage of minced veal, sealed in plastic bags, for 2 weeks at 4 °C.[83] Lamb slices can be converted into a shelf-stable product of good texture and flavour, if heated to 71–76 °C, followed by freeze desiccation for 90 min at −19 °C, vacuum packing and irradiation at 5–10 kGy at −30 °C. The product could be stored for 90 days at 21–25 °C.[84]

7. WHOLESOMENESS ASPECTS OF MEAT IRRADIATION

7.1. General Considerations

The maximum energies of gamma rays emitted by ^{60}Co or ^{137}Cs do not induce radioactivity in irradiated foods, neither do treatment by electron rays at 10 MeV or X-rays at 5 MeV. This aspect is therefore no longer relevant for the wholesomeness evaluation of irradiated meats. Thorough consideration of the microbiological aspects has also not furnished any practical evidence of hazard from pathogens, or toxin-producing micro-organisms. The nutritional adequacy of radurised, radicidised or radappertised foods has been demonstrated by the absence of significant impairment of the proteins, lipids or carbohydrates in the food. There is also the additional protection from the reduced sensitivity to irradiation of the nutritional components in the actual food compared to the effects of irradiating pure solutions or artificial mixtures. Low temperatures and the absence of oxygen further minimise the effects of the radicals generated by the radiation through their immobilisation and the reduced shielding effect of ice.

Studies of the radiolytic changes in irradiated foods originally served to define the optimal dose, temperature and other conditions for achieving a satisfactory technological result. Other objectives concerned the possible changes in nutritional value and the search for radiolytic products which might serve as marker substances for the identification of irradiated foods.

The nature and quantities of radiolytic products generated in foodstuffs subjected to irradiation are important determinants in the evaluation of the safety to health of foodstuffs processed by this technology. There was therefore a considerable impetus to discover common features among the radiolytic products produced in individual foods to be able to extrapolate findings between members of a given class of foods and also between classes of foods. Such knowledge was pivotal in changing the toxicological appraisal of the wholesomeness of irradiated foods from the unsatisfactory demand for more feeding experiments in several animal species to an appreciation of the nature and extent of radiolytic effects in irradiated foods. The concept of combining knowledge of the radiation chemistry of food components with judiciously selected animal feeding studies and genotoxic investigations led eventually to a more rational appraisal of the wholesomeness of irradiated foods.

Further evidence was provided by the demonstration of the palatability and nutritional adequacy of animal feeds radappertised with 25 kGy through the raising of germ-free rodent colonies over many years.[140] Composite diets for patients lacking immunological defences have been radappertised with 25 kGy at −15 °C and were found to be wholesome.[140] Radappertised meats had also been tested in feeding studies in rats and dogs using *in utero* exposure followed by exposure throughout life. Rodents were studied over four generations and dogs for 3 years while they produced three litters. The usual biochemical and pathological and histopathological parameters were observed, although it should be realised that traditional toxicological approaches were not strictly applicable to the testing of foods.[119]

Supplementary evidence was provided by the toxicological evaluation of identified radiolysis products in treated foods. These occurred at concentrations ranging from 1–100 μg/kg beef up to a total of 9·4 mg/kg. The bulk of the volatile compounds arose from the fat moiety of the meat, the less volatile components being identified by size exclusion chromatography as pentadecanoyl and (9, 10-ene)heptadecanoyl radicals, which recombine to form long-chain hydrocarbons, ketones and diketones.[122,123] None of these compounds were found to present a toxicological hazard at the concentrations detected.[113] This view was supported by the uniformity and predictability of the chemical changes caused by irradiation and the lack of genotoxicity.[17] Moreover this predictability of the yields of major radiolytic compounds led to the internationally accepted view that no further animal testing or toxicological testing was necessary for foods irradiated with doses up to an average overall dose of 10 kGy.[115] Further studies have shown that, over the range of 10–71 kGy applied to foods frozen at −40 to −20 °C, the radiolysis products generated were qualitatively similar to those found with lower doses at ambient temperature but quantitatively lower.

7.2. Beef and Beef Products

A number of short-term feeding tests in rats, extending over 8–12 weeks, have been performed with 35 % irradiated beef in the diet to explore the effects of storing the treated beef for 3–6 months at 70° and 100°F, and of the consumption of raw and cooked irradiated beef. No evidence of toxicity was noted.[107]

A large number of long-term feeding tests have also been carried out in rats and dogs using beef or beef products as the major irradiated component in the diet. The dietary levels were 35–45% of the total dietary solids. Feeding extended mostly over 2 years. Minced beef was most commonly used but irradiated beef stew was also employed. Radiation sources, packaging, storage periods and storage temperatures varied in these experiments but 28 kGy and 56 kGy were the most common dose levels employed. No consistent evidence of toxicity or carcinogenicity was found. Early studies were marred by the effects of vitamin E deficiency on the reproductive phases.

Some lifespan studies in rats used irradiated cooked beef, others used canned beef stew stored for 3–9 months before incorporation in the diet. No consistent adverse effects were noted in the parent and four filial generations.[82,118]

Several 2-year studies in dogs used minced beef, irradiated frozen with 28 or 56 kGy, at 35% of the diet. Dogs on the test diets conceived less well than the controls, had smaller litters, and pup mortality was slightly greater. No macroscopic or histopathological abnormalities were found but the small number of animals used made interpretation difficult.[124,125] Another 3-year study in beagles with minced beef, irradiated at 60 kGy, showed no abnormal findings and normal reproduction.[126] Feeding cooked irradiated meat, stored for 36 weeks at ambient temperature, to beagles for 2 years confirmed the absence of deleterious effects.[127] Similar negative results were noted in an 18-month study in dogs fed beef, irradiated at 6 kGy, at a dose level of 200 g/day.[128]

Dominant lethal tests with vacuum packed raw meat, treated with 6 kGy and then cooked, and also with roast meat treated with 8 kGy were negative in rats at doses of 10 g/animal.[133] Beef irradiated with 47–71 kGy and stored for 10 weeks did not induce genetic aberrations in Drosophila.[131] Pan-fried minced low and high fat beef (12 min at 230 °C) gave a positive result with standard five tester strains of *Salm. typhimurium*. Non-irradiated meat gave the same result as meat irradiated with 50 kGy or cooked irradiated meat, tested immediately after irradiation, or stored for 12 weeks at 20 °C in polythene bags. The addition of cereal extender or single-cell protein reduced the observed mutagenicity by a dilution effect. Frying with butter or maize oil also reduced the mutagenic activity.[75]

7.3. Pork

No adverse effects on the growth of rats were noted when fed for 12 weeks pork, irradiated at 28 and 56 kGy and stored for 3–6 months, at 35 % of the diet. Liver xanthine oxidase and alkaline phosphatase activity were unaffected but cytochrome oxidase activity was increased in proportion to the radiation dose.[93,107]

Rats were also fed for 2 years minced, can-sealed pork, irradiated at 28 and 56 kGy and stored at room temperature for 3–8 months, cooked prior to incorporation in the diet at 35 %. Four generations were produced. No treatment-related effects were noted.[94,95]

Pork, irradiated at 28 and 56 kGy and stored for 3 months at 21 °C, was cooked and fed at 35 % of the diet to beagles for 2 years without adverse effects or impairment of reproduction.[124] Similarly pork, treated with 8 kGy, was fed to beagles at 200 g/animal/day without causing any adverse effects in 18 months.[128] Fresh minced pork, irradiated at 26 kGy and stored in plastic pouches at 15–35 °C for 30 days, was fed to chow dogs for 4 years at 35 % of the diet without ill effects or adverse changes in bone marrow cytogenetics and in the cytogenetics of peripheral lymphocytes.[100]

7.4. Ham

Cooked ham, irradiated with 20 kGy and stored at −12 °C, was fed to rats for 35 weeks at 41·5 % of the diet. A comparison group was fed ham irradiated with 2·5 kGy. No adverse effects were noted.[99]

In a lifespan study in rats ham, irradiated at 20 kGy, was fed at 25 % of the diet. No abnormal findings were noted nor was reproduction affected.[99] Another lifespan study in mice, using similarly treated ham at 24 % of the diet, produced no abnormal effects.[99]

A complex feeding study in rats used autoclaved or irradiated cured pork at 35 % of the diet for 2·5 years. The meat came from pigs reared on three different diets, e.g. untreated, irradiated with 50 kGy or autoclaved at 120 °C for 10 min, and subsequently cured with brine containing 50 or 200 mg/kg nitrite. The cured meat was either pasteurised at 75 °C, irradiated at 37 kGy or 74 kGy at −30 °C, or autoclaved at 110 °C. The six test groups included groups with high nitrite and with low nitrite combined with low or high dose irradiation. No significant treatment-related effects were noted nor was there any evidence of carcinogenicity.[101]

Ham, irradiated at 37–42 kGy and stored for 10 weeks, did not induce any genetic aberrations in Drosophila.[131]

7.5. Bacon

Rats were fed for 8–12 weeks variously treated bacon at 35% of the diet without showing any treatment-related effects apart from some increase in hepatic and erythrocyte cytochrome oxidase. In these studies cured and smoked canned bacon was irradiated by either machine or isotope sources at doses of 45–56 kGy and subsequently kept either at room temperature or under refrigeration for periods from 1 month to 1 year.[146]

Young healthy male volunteers consumed mixed diets of meat, fruit, vegetables and cereals for 15 days. Irradiated foods were incorporated at 35, 60, 80 and 100% of the calories. Bacon sterilised with 37 kGy was included in these trials. Clinical chemistry, urinalysis and digestibility of irradiated and control bacon were similar.[147]

Beagles were fed cooked bacon, irradiated at 28 and 56 kGy, and subsequently stored 6 months at ambient temperature, at 35% of the diet for 2 years. Reproduction was poor but no other adverse effects were noted.[148]

Rats were fed bacon, irradiated at 28 and 56 kGy, stored for 3 months at room temperature and then cooked, at 35% of the diet in a factorial design with fruit compote. Mortality was slightly increased in groups fed bacon and/or fruit irradiated at 56 kGy. All other parameters including cholesterol/phospholipid ratios of brain lipids were unaffected by irradiated bacon. Reproduction of three filial generations showed some reduced live births and number of pups weaned and some increase in still births in all groups fed irradiated bacon. However the small numbers made interpretation difficult.[149]

A carcinogenicity study in two strains of mice with total lipids from bacon irradiated at 56 kGy and stored for 6 months, fed at 10–20% of the diet, revealed no tumorigenic effect.[150]

Bacon irradiated at doses of 45–56 kGy was approved by the US Food and Drug administration in 1963. This was rescinded in 1968 because the data were considered inadequate for the more stringent requirements then operative.

7.6. Poultry Meat

Two studies in rats, extending over 2 years and incorporating a multi-generation reproduction phase, used either fresh boned chicken irradiated with 28 or 56 kGy and beans irradiated similarly or chicken stew and cabbage irradiated at similar doses. The irradiated foods were stored for 3–6 months before incorporation in the diet at 35% in a factorial design. Neither chicken and beans nor chicken stew and cabbage produced any adverse effects. Enzyme levels of the intestinal mucosa showed no consistent changes.[152,153]

Beagles were fed irradiated chicken stew (28 or 56 kGy) stored at room temperature for 3–6 months, at 35 % of the diet for 2 years without adverse effects.[124]

A large complex study in mice, rats and beagles, using *in utero* exposure and extending over 2 years in mice and rats and 3 years in dogs, was carried out with chicken either sterilised at 45–59 kGy by gamma or electron radiation, or enzyme-inactivated frozen or thermally sterilised. Results were compared to controls on a basal diet. Irradiated chicken was precooked and vacuum sealed. The level of dietary incorporation was 35 %. Oncogenicity was determined on the *in utero* exposed generation. The reproductive phase included three filial generations. Contradictory interpretations of the findings have been reported and these differences have not yet been resolved. The heavier F_0 males and the female mice fed gamma-irradiated chicken had apparently higher mortalities in the second year but the relevant group sizes were small. Reproduction data were satisfactory for all test groups. The F_0 group fed gamma-irradiated chicken showed a higher tumour incidence than other groups but the tumoric incidence varied with sex and irradiation type. Female mice fed gamma-irradiated chicken had the highest incidence of alveogenic tumours but there were more deaths in the gamma-irradiated group, making interpretation difficult. Glomerulonephritis had an earlier onset in the gamma-irradiated group but incidence and severity were similar in all test groups. Thus no clear findings on the effect of feeding irradiated chicken to mice were obtained.

The rat study was abandoned in the 39th week due to high neonatal mortality in all groups in the F_{1b} generation. At that time no adverse effects were noted. The dog study showed no treatment-related deleterious effects.

Teratology studies in rabbits using 35 % and 70 % irradiated chicken in the diet revealed no teratogenic effects. Similar tests in mice, hamsters and rats were also negative.

The same irradiated chicken meats were examined for mutagenic potential. The Salmonella reverse mutation test was negative. The heritable translocation test in mice showed no evidence of chromosomal damage. A dominant lethal test in mice was negative when meats were fed at 35 % of the diet. Because the positive control group also showed no dominant lethal effect the test results have to be discarded. A sex-linked recessive lethal test in *Drosophila* produced no mutations but a reduced number of offspring in the gamma-irradiated chicken group. The effect was dose-related. Frozen chicken gave a similar though smaller response.[154]

Chicken meat irradiated at 7 kGy in ambient temperature was cooked and the aqueous extract as well as an in-vitro digest tested for mutagenicity in the Salmonella reverse mutation test, in cultured Chinese hamster ovary cells for induced forward mutations and sister chromatid exchanges and *in vivo* in mice, rats and hamsters, fed cooked irradiated chicken, by the micronucleus test, by bone marrow cytogenetics and for sister chromatid exchanges. No mutagenic activity was detected.[158]

7.7. Mixed Offal

A 2-year four generation feeding test was carried out in rats with a mixture of beef kidney, heart, spleen, spinal cord, brain, tripe and pancreas. Offals were frozen, irradiated at 28 kGy and stored at −20 °C, and fed at 60 % in the diet. No adverse effects were noted.[157]

7.8. Composite Diets Containing Meat

Life span and multigeneration tests in rats and mice using a variety of mixed diets irradiated at doses from 1 to 56 to 100 kGy and stored for 3 to 9 months, showed no adverse effects of significance.[79,80,81,159]

Two-year studies in beagles with mixed diets, irradiated at 28 or 56 kGy and stored for 3–12 months at room temperature, gave no adverse results on feeding at 35 % of the diet.[109]

Nine human subjects were given for 15 days irradiated or non-irradiated mixed diets in which 35 % of the calories were irradiated foods, treated at 30 kGy. No clinical abnormalities were noted.[76]

7.9. Horse Meat

Rats were fed for 18–21 months, at 50 % of the diet, frozen horse meat, irradiated at −15 °C with 6·5 kGy and stored for 2–6 weeks before being cooked for incorporation in the rodent diet. One generation was produced. No adverse effects were noted and reproduction was comparable to controls.[85]

8. COST AND ENERGY CONSIDERATIONS

Legal clearance of the process of food irradiation does not necessarily lead to its commercial application. Only if the costs are competitive with alternative processes and if the benefits are commensurate, will there be a developing industrial interest in exploiting the technology. Many factors such as the nature of the product, the location and capacity of the irradiation plant, the accessibility of the facility, the source characteristics and the capital investment required enter the cost–benefit calculations. For example, for irradiation sterilisation at 26 kGy of refrigerated bacon at an annual throughput of 100 million pounds the cost might range from 0·76–3·23 cents/lb. If irradiation were to be carried out on frozen bacon at −40 to −20 °C, the cost of liquid nitrogen would raise the cost to 10 cents/lb at an annual throughput of 1 million pounds or 1·94 cents/lb at 30 million pounds annual throughput.[116]

The energy required for radappertisation is about 157 kJ/kg food processed.[117] The energy saving for irradiated products versus frozen products is about 45 %.[102]

9. REFERENCES

1. Wüst, O. (1930). French Pat. 701302, cit. *Chem. Abstr.*, **25**, 4068.
2. Brasch, A. and Huber, W. (1947). *Science*, **105**, 112.
3. Goresline, H. E. (1973). In: *Radiation preservation of food*, IAEA Proceedings Series, Vienna, p. 1.
4. Diehl, J.-F. (1983). In: *Preservation of food by ionizing radiation* (Josephson, E. S. (Ed.)), Vol. I, CRC Press Inc., Boca Raton, Florida, p. 279.
5. Nawar, W. W. (1977). In: *Radiation chèmistry of major food components* (Elias, P. S. and Cohen, A. J. (Eds)), Elsevier, Amsterdam, Chap. 3.
6. Bailey, A. J. and Rhodes, D. N. (1964). *J. Sci. Food Agric.*, **15**, 504.
7. Morgan, B. H. and Sin, R. G. H. (1957). In: *Radiation preservation of food*, US Army Quartermaster Corps., US Government Printing Office, Washington DC, Chap. 26.
8. El-Zeany, B. A., Abdel-Fattah, L. E. and Hassan, I. M. (1980). *Z. Lebensm.-Unters. Forsch.*, **171**, 5.
9. Josephson, E. S., Brynjolfsson, A., Wierbicki, E., Rowley, D. B., Merritt, C. Jr, Baker, R. W., Killoran, J. J. and Thomas, M. H. (1973). In: *Radiation preservation of food.* IAEA STI/PUB/317, Vienna, p. 471.
10. Hannan, R. S. and Shepherd, H. J. (1959). *J. Sci. Food Agric.*, **10**, 286.
11. Coleby, B., Ingram, M., Shepherd, H. J., Thornley, M. J. and Wilson, G. M. (1961). *J. Sci. Food Agric.*, **12**, 483.
12. Coleby, B., Ingram, M. and Shepherd, H. J. (1961). *J. Sci. Food Agric.*, **12**, 417.
13. Wilson, G. M. (1959). *J. Sci. Food Agric.*, **10**, 295.
14. Sheffner, A. L. and Spector, H. (1957). In: *Radiation preservation of food*, US Army Quartermaster Corps., US Government Printing Office, Washington DC, Chap. 18.
15. Diehl, J.-F. (1979). *Z. Lebensm.-Unters. Forsch.*, **169**, 276.

16. Diehl, J.-F. (1979). *Z. Lebensm.-Unters. Forsch*, **168**, 29.
17. Taub, I. A., Robbins, F. M., Simic, M. G., Walker, J. E. and Wierbicki, E. (1979). *Food Technol. (Chicago)*, **33**(5), 184.
18. Gernon, G. D. Jr, Kraus, F. J. and Drake, M. P. (1961). *Food Technol. (Chicago)*, **15**, 354.
19. Diehl, J.-F. (1969). *Food Irrad.*, **10**(1/2), 2.
20. Wick, E. L., Koshika, M. and Mizutani, J. (1965). *J. Food Sci.*, **30**, 433.
21. Wills, E. D. (1980). *Int. J. Radiat. Biol.*, **37**, 403.
22. Batzer, O. F. and Doty, D. M. (1955). *J. Agric. Food Chem.*, **3**, 64.
23. Sribney, M., Lewis, U. J. and Schweigert, B. S. (1955). *J. Agric. Food Chem.*, **3**, 958.
24. Marbach, E. P. and Doty, D. M. (1956). *J. Agric. Food Chem.*, **4**, 881.
25. Batzer, O. F., Sribney, M., Doty, D. M. and Schweigert, B. S. (1957). *J. Agric. Food Chem.*, **5**, 700.
26. Pearson, A. M., Dawson, L. E., Bratzler, L. J. and Costilow, R. N. (1958). *Food Technol. (Chicago)*, **12**, 616.
27. Hedin, P. A., Kurtz, G. W. and Koch, R. B. (1961). *Food Res.*, **25**, 382.
28. Ingram, M., Coleby, B., Thornley, M. J. and Wilson, G. M. (1959). *Proc. Int. Conf. Preserv. Foods Ionizing Radiation. Dept. Food Technol.*, MIT, Cambridge, Mass., p. 161.
29. Batzer, O. F., Sliwinski, R. A., Chang, L., Pih, K., Fox, J. B. Jr, Doty, D. M., Pearson, A. M. and Spooner, M. E. (1959). *Food Technol. (Chicago)*, **13**, 501.
30. Martin, S., Batzer, O. F., Landmann, W. A. and Schweigert, B. S. (1962). *J. Agric. Food Chem.*, **10**, 91.
31. Burks, R. E. Jr, Baker, E. B., Clark, P., Esslinger, J. and Lacey, J. C. Jr (1959). *J. Agric. Food Chem.*, **7**, 778.
32. Monty, K. J., Tappel, A. L. and Groninger, H. S. (1961). *J. Agric. Food Chem.*, **9**, 55.
33. Merritt, C. Jr (1972). *Radiat. Res. Rev.*, **3**, 353.
34. Vajdi, M., Nawar, W. W. and Merritt, C. Jr (1979). *J. Am. Oil Chem. Soc.*, **56**, 611.
35. Vajdi, M., Nawar, W. W. and Merritt, C. Jr (1979). *J. Am. Oil Chem. Soc.*, **56**, 906.
36. Partmann, W. and Schlaszus, H. (1980). *Z. Lebensm.-Unters. Forsch.*, **171**, 1.
37. Nawar, W. W. and Balboni, J. J. (1970). *J. Am. Oil Chem. Soc.*, **53**, 726.
38. Merritt, C. Jr, Angelini, P. and McAdoo, D. J. (1961). In: *Radiation preservation of foods, Adv. Chem. Ser. 65*, Amer. Chem. Soc., Washington DC, Chap. 3.
39. Wick, E. L., Murray, E., Mitzutani, J. and Koshika, M. (1967). In: *Radiation preservation of foods, Adv. Chem. Ser. 65*, Amer. Chem. Soc., Washington DC, Chap. 2.
40. Soják, L., Hrivnák, J. and Salková, Z. (1974). *Radiochem. Radioanal. Litt.*, **20**, 87.
41. Drawert, F. and Beck, B. (1972). *Chem. Mikrobiol. Technol. Lebens.*, **1**, 110.
42. Drawert, F. and Beck, B. (1974). *Z. Lebens.-Unters. Forsch.*, **155**, 282.

43. LICCIARDELLO, J. J., NICKERSON, J. T. R., PROCTOR, B. E. and CAMPBELL, C. L. (1959). *Food Technol. (Chicago)*, **13**, 398 and 405.
44. GREENE, B. E. and WATTS, B. M. (1966). *Food Technol. (Chicago)*, **20**(8), 111.
45. LAWRIE, R. A., SHARP, J. G., BENDALL, J. R. and COLEBY, B. (1961). *J. Sci. Food Agric.*, **12**, 742.
46. SHAPIRA, R. (1963). *Int. J. Radiat. Biol.*, **7**, 537.
47. HAMM, R. and DEATHERAGE, F. E. (1960). *Food Res.*, **25**, 587.
48. WHITING, R. C. and RICHARDS, J. F. (1971). *J. Food Sci.*, **36**, 752.
49. UZUNOV, G., TSOLUVA, C. and NESTOROV, N. (1972). *Int. J. Radiat. Biol.*, **22**, 437.
50. SATTERLEE, L. D., BROWN, W. D. and LYCOMETROS, C. (1972). *J. Food Sci.*, **37**, 213.
51. ALTMANN, H., KLEIN, W. and DOLEJS, I. (1974). In: *The identification of irradiated foodstuffs*, Commission of the European Communities, EUR 5126, Luxembourg, p. 61.
52. DESAI, I. D., SAWANT, D. L. and TAPPEL, A. L. (1964). *Biochim. Biophys. Acta*, **86**, 277.
53. KLEIN, W. and ALTMANN, H. (1975). *Kerntechnik*, **14**, 582.
54. MCGOWN, E. L., LEWIS, C. M. and WARING, P. P. (1979). Letterman Army Institute of Research Report No. 71, San Francisco, USA.
55. MCGOWN, E. L., LEWIS, C. M. and WARING, P. P. (1979). Letterman Army Institute of Research Report No. 72, San Francisco, USA.
56. MAXCY, R. B. and ROWLEY, D. B. (1978). In: *Food preservation by irradiation*, Vol. I (Proc. Int. Symp. Wageningen, Nov. 1977) IAEA, p. 347.
57. WELCH, A. B. and MAXCY, R. B. (1979). *J. Food Sci.*, **44**, 673.
58. WELCH, A. B. and MAXCY, R. B. (1975). *Appl. Microb.*, **30**, 242.
59. SNYDER, L. D. and MAXCY, R. B. (1979). *J. Food Sci.*, **44**, 33.
60. MAHNEL, R., STETTMUND VON BRODOROTTI, H. and OTTIS, K. (1980). *Zentbl. Bakt. 1. Abt. Orig. B*, **170**, 57.
61. TEUFEL, P. (1983). In: *Recent advances in food irradiation* (Cohen, A. J. and Elias, P. S. (Eds)), Elsevier Biomedical Press, Amsterdam, p. 217.
62. LAPIDOT, M. (1979). *Radiat. Phys. Chem.*, **14**, 671.
63. GOMBERG, H. J. and GOULD, S. E. (1983). *Science*, **118**, 75.
64. POWEL, O. (1968). *Devitalization of cysticerci by Gamma Radiation. Elimination of Harmful Organisms from Food and Feed by Irradiation*, Proceedings of a Panel, Zeist, 12-16/6/67, IAEA, Vienna.
65. GELFAND, S. YU. and ANISIMOV, B. N. (1973). *IAEA Symposium*, Sofia, 15-17/10/73.
66. SHULTS, G. W., COHEN, J. S. and WIERBICKI, E. (1975). Radiation inactivation of meat proteases as determined by a C^{14}-labelled hemoglobin method. Natick Report 76-33 FEL.
67. WOLF, A. (1968). *Vitalst-Zivil.*, 2nd April, 1968.
68. MERRITT, C. JR, ANGELINI, P., WIERBICKI, E. and SHULTS, G. W. (1975). *J. Agric. Food Chem.*, **23**(6), 1037.
69. YOUSRI, R. M. (1980). *Z. Ernährungswiss.*, **19**, 111.
70. DELINCÉE, H. (1983). In: *Recent advances in food irradiation* (Cohen, A. J. and Elias, P. S. (Eds)), Elsevier Biomedical Press, Amsterdam, p. 129.
71. RANIERI, S. (1979). *Food Prod. Dev.*, **13**(10), 28.

72. Simic, M. G., Dizdaroglu, M. and De Graff, E. (1982). *4th Int. Meeting Rad. Process*, Dubrovnik.
73. Maxcy, R. B. (1973). Study of Control of public health problems using irradiation. Final report Coo-2038-4.
74. Firstenberg-Eden, R., Rowley, D. B. and Shattuck, E. (1980). *Appl. Env. Microb.*, **39**(1), 159.
75. Münnzer, R. (1983). *Fleischwirts*, **63**(4), 611.
76. McGary, V. E., Shipman, G. E. and Bernskin, L. M. (1956). *J. Amer. Diet. Assoc.*, **32**, 123.
77. Stockhausen, K. and Bögl, W. (1978). In: *Food preservation by irradiation*, Vol. I, IAEA-SM-221/19, p. 421.
78. Josephson, E. S., Thomas, M. H. and Calhoun, W. R. (1976). *First Int. Congr. Eng. and Food, Boston*, 9–13th Aug., 1976.
79. Kline, B. E. and Teply, L. J. (1959). The possible carcinogenicity of irradiated foods, Final Report OGS contract DA-49-007 MD-583, Dec, 1959.
80. Calandra, J. C. and Kay, J. H. (1963). The carcinogenic properties of irradiated foods, Final Report OGS contract DA-49-007-MD-895. Feb, 1963.
81. Radomski, J. L., Deichmann, W. B., Austin, B. S., MacDonald, Wm. E. and Barnal, E. (1965). *Tox. Appl. Pharm.*, **7**, 122.
82. Blood, F. R., Wright, M. S., Darby, W. J. and Elliott, G. A. (1966). *Toxicol. Appl. Pharm.*, **8**. 241.
83. Grünewald, Th. and Münzner, R. (1973). *Fleischwirts.*, **53**, 391.
84. Rao, B. Y. K., Shults, G. W. and Wierbicki, E. (1974). Preservation of enzyme-inactivated lamb slices—effects of freeze desiccation and low dose irradiation, Techn. Rep. 74-34-FL, Natick Labs.
85. Hickman, J. R., Law, A. W. and Ley, F. J. (1969). Studies on the wholesomeness of irradiated meat, AERE Report R 6025, Wantage, UK.
86. Marito, M., Tajima, M. and Fujimaki, M. (1971). *Agr. Biol. Chem.*, **35**(12), 1936.
87. De Groot, A. P., van der Mull Dekker, L. P., Slump, P., Vos, H. J. and Willems, J. J. L. (1972). Report R 3787, CIVO, Zeist, Netherlands.
88. Harmuth-Hoene, A.-E. and Delincée, H. (1978). *Int. J. Vit. Nutr. Res.*, **48**, 62.
89. Yousri, R. M. and Harmuth-Hoene, A.-E. (1979). *Int. J. Vit. Nutr. Res.*, **49**, 171.
90. Harmuth-Hoene, A.-E. (1976). *Int. J. Vit. Nutr. Res.*, **46**, 348.
91. Dimitrova, N., Grozdanov, A., Dikova, G., Gadjeva, D., Tencheva, S., Katsareva, T. and Shopova, E. (1980). *Biological evaluation of actinomysin–lard mixtures—effect of irradiation and subsequent storage*. IAEA-TECDOC-256, Vienna, p. 139.
92. Lawrie, R. A. (1968). *J. Sci. Food Agric.*, **19**, 233.
93. Read, M. S., Trabosh, H. M. and Kraybill, H. F. (1959). *Fed. Proc.*, **18**, 542.
94. Buble, E. C. and Butts, J. S. (1960). *J. Nutr.*, **70**, 211.
95. Tinsley, I. J., Buble, E. C., Butts, J. S. and Bone, J. F. (1964). The growth, breeding, longevity and histopathology of rats fed irradiated or control foods, Final Report OSG contract DA-49-007-MD-580, Sept., 1964.
96. Anellis, A., Berkowitz, D., Jarbol, C. and El-Bisi, H. M. (1967). *Appl. Microb.*, **15**(1), 166.

97. WIERBICKI, E. and HEILIGMAN, F. (1973). *Proc. Int. Symp. Nitrite Meat Prod.*, Zeist, p. 189.
98. ANELLIS, A., SHATTUCK, E., LATT., T., SONGPASERTCHAI, S., ROWLEY, D. B. and ROSS, E. W. JR (1976). *Ann. Meetg Amer. Soc. Microb.*, **76**, 189.
99. JENKINS, F. P. (1962). *Proc. FAO Techn. Meet. Eval. Wholesomeness Irrad. Foods*, Brussels, Oct, 1961, p. 136.
100. KE-WEN, C. and DAO-JING, Z. (1983). *Radiat. Phys. Chem.*, **22**(3), 792.
101. VAN LOGTEN, M. J., DE VRIES, T., VAN DER HEIJDEN, C. A., VAN LEEUWEN, E. X. R., GARBIS-BERKVENS, MW. J. M. and STRIK, J. J. T. W. A. (1983). Nat. Inst. Publ. Health, Netherlands, Report 61740001.
102. BRYNJOLFSSON, A. (1979). Cost of irradiating bacon and the associated energy savings, US Army Natick R & D Command, Techn. Rep. TR-79/022, March, 1979.
103. HAMM, R., HOFMANN, K., GRÜNEWALD, TH. and PARTMANN, W. (1975). Veränderungen von Aminotransferasen und Muskelproteinen bei der Behandlung von Schweinefleisch mit ionisierenden Strahlen, Private Communication.
104. KAMAREI, A. R., KAREL, M. and WIERBICKI, E. (1981). *J. Food Sci.*, **46**, 37.
105. URBAIN, W. M. (1983). In: *Preservation of food by ionizing radiation* (Josephson, E. S. and Peterson, M. S. (Eds)), Vol. III, CRC Press, Boca Raton. Florida, USA, p. 2.
106. SUDARMADJI, S. and URBAIN, W. M. (1972). *J. Food Sci.*, **37**, 671.
107. READ, M. S., TRABOSH, H. M., WORTH, W. S. and KRAYBILL, H. F. (1959). *Toxicol. Appl. Pharm.*, **1**, 417.
108. POLING, E. C., WARNER, W. D., HUMBURG, F. R., REPER, E. F., URBAIN, W. M. and RICE, E. E. (1955). Growth reproduction, survival and histopathology of rats fed beef irradiated with electrons.
109. BLOOD, F. R., DARBY, W. J., ELLIOT, G. K. and WRIGHT, M. S. (1966). *Toxicol. Appl. Pharm.*, **8**, 235.
110. ANELLIS, A., SHATTUCK, E., ROWLEY, D. B., ROSS, E. W. JR, WHALEY, D. N. and DOWELL, V. R. JR (1975). *Appl. Microb.*, **30**(5), 811.
111. KAUFFMAN, F. L. and HARLAN, J. W. (1969). Effect of low temperature irradiation on chemical and sensory characteristics of beefsteaks, Techn. Rep. 69-64-FL.
112. KUNG, H., GADEN, E. L. and KING, C. G. (1953). *J. Agric. Food Chem.*, **1**, 142.
113. FASEB (1977 and 1979). Select committee on health aspects of irradiated beef, Contract No. DAMD-17-76-C-6055; Bethesda, Md, August 1977, Suppl. I & II, March 1979.
114. TAUB, I. A., KAPRIELIAN, R. A., HALLIDAY, J. W., WALKER, J. E., ANGELINI, P. and MERRITT, C. JR (1979). *Radiat. Phys. Chem.*, **14**, 639.
115. FAO IAEA/WHO (1981). Report of Joint Expert Committee on wholesomeness of irradiated food. WHO-Techn. Rep. Ser. 659, WHO, Geneva.
116. JOSEPHSON, E. S. (1983). In: *Preservation of food by ionizing radiation* (Josephson, E. S. and Peterson, M. S. (Eds)). Vol. III, CRC Press Inc., Boca Raton, Florida, p. 246.
117. BRYNJOLFSSON, A. (1978). In: *Food preservation by irradiation*, Vol. 2, STI/PUB/470, IAEA, Vienna, p. 285.

118. RADOMSKI, J. L., DEICHMANN, W. B., AUSTIN, B. S. and MACDONALD, W. E. (1965). *Toxicol. Appl. Pharm.*, **7**, 113.
119. RAICA, N. JR and BAKER, R. W. (1972). The wholesomeness testing of radappertized enzyme-inactivated beef, Report to IAEA 166/45.
120. UZUNOV, G., Tsolova, L. and NESTOROV, N. (1972). *Int. J. Radiat. Biol.*, **22**(5), 437.
121. WIERBICKI, E. (1980). *Proc. 26th Europ. Meetg. Meat Res. Workers, Colorado Springs*, 31/8–5/9/1980.
122. VAJDI, M. and NAWAR, W. W. (1979). *J. Amer. Oil Chem. Soc.*, **56**, 611.
123. VAJDI, M., NAWAR, W. W. and MERRITT, C. JR (1979). *J. Amer. Oil Chem. Soc.*, **56**, 906.
124. MCCAY, C. M. and RUMSEY, G. L. (1960). *Fed. Proc.*, **19**, 1027.
125. LOOSLI, J. K., MCCAY, C. M., STEVENS, A. C. and KENNY, J. W. (1964). Components of ionized irradiated meats injurious to reproduction, Final Report, June, 1964.
126. CLARKSON, T. B. and PICK, C. R. (1964). The effect of control-ground beef and irradiated 5·58 Mrad consumption on reproductive performance in the beagle, Final Report, Sept., 1964.
127. REBER, E. R., MALHOTRA, O. P., KREIER, J. P., NORTON, H. W. and BEAMER, P. D. (1962). *Amer. J. Vet. Res.*, **23**, 74.
128. SHILLINGER, YU-I., KACHKOVA, V. O. and MAGANOVA, N. B. (1965). *Voprosy Pitaniya*, **24**(1), 19.
129. URBAIN, W. M., GIDDINGS, G. G. and BALLANTYNE, W. W. (1972). *Isotope & Rad. Techn.*, **9**(3), 288.
130. MACFARLANE, J. J., EUSTACE, I. J. and GRAU, F. H. (1983). *Proc. Nat. Symp. on Ionizing Energy Treatment of Foods, Sydney*, 5/10/1983, p. 39.
131. MITTLER, S. (1979). *Int. J. Radiat. Biol.*, **35**(6), 583.
132. GELFAND, S. YU., NOMEROTSKAYA, N. F. and MOZUL, N. YA. (1973). *CMEA Symposium: Radiation Treatment of Food and Agricultural Products, Sofia*, 15–17/10/1973.
133. KAMALDINOVA, Z. M., SHILLINGER, YU. I. and ZAJTSEV, A. N. (1977). *Vopros. Pit.*, March–April, 53.
134. WIERBICKI, E. (1980). US Army Natick R & D Laboratories, Natick, Mass., Personal communication, 31/7/1980.
135. SHULTS, G. W., RUSSELL, D. H. and WIERBICKI, E. (1972). *J. Food Sci.*, **37**, 860.
136. FAO/IAEA/WHO (1964). Joint Expert Committee Report, The technical basis for legislation on irradiated food, WHO Techn. Rep. Ser. No. 316.
137. ANELLIS, A. and KOCH, R. B. (1962). *Appl. Microbiol.*, **10**, 326.
138. KILLORAN, J. J. (1972). *Radiat. Res. Rev.*, **3**, 369.
139. WIERBICKI, E. and BRYNJOLFSSON, A. (1979). *Proc. 25th Eur. Meat Res. Workers, Budapest*, 27–31/10/1979.
140. LEY, F. J. (1975). *Proc. 5th Int. Cong. Radiat. Res., Seattle, 1974* (Nygaard, O. F., Adler, H. I. and Sinclair, W. K. (Eds)), Acad. Press, NY, p. 118.
141. PLOUGH, I. C. (1957). An evaluation in human beings of the acceptability, digestibility and toxicity of pork sterilized by gamma radiation and stored at

room temperature, Rep. No. 204. US Army Med. Nutr. Lab., Denver, Colorado.
142. Schreiber, M. and Nasset, E. S. (1959). *J. Appl. Physiol.*, **14**, 639.
143. Ley, F. J., Bleby, J., Coats, M. E. and Batterson, J. S. (1969). *Lab. Anim.*, **3**, 221.
144. Metlitsky, L. V., Rogachev, V. N. and Krushchev, V. G. (1968). In: *Status of the food irradiation program*. Congress of US, Washington, DC, p. 30.
145. Sheffner, A. L. and Spector, H. (1957). In: *Radiation preservation of food*, US Army Quartermaster Corps, Washington, chap. 18.
146. Read, M. S., Kraybill, H. F. and Witt, N. F. (1958). *J. Nutr.*, **65**, 39.
147. Plough, I. C., Bierman, E. I., Levy, L. M. and Witt, N. F. (1969). *Fed. Proc.*, **19**, 1952.
148. Hale, M. W., Schroeder, W. F. and Sikes, D. (1960). Growth, reproduction, mortality and pathological changes in dogs fed gamma-irradiated bacon for 2 years. Final report OSG contract DA-49-007-MD-780, March 1960.
149. Mead, J. F. and Griffith, W. H. (1959). Effect of ionizing radiation on the nutritive and safety characteristics of food. Final Report OSG contract DA-49-007-MD-579.September 1959.
150. Dixon, M. S., Mayer, D. L., Zeldis, L. J. and McKee, R. W. (1961). *J. Food Sci.*, **26**, 611.
151. Fiddler, W., Gates, R. A., Pensabene, J. W., Phillips, J. G. and Wierbicki, E. (1981). *J. Agric. Food Chem.*, **29**, 551.
152. Richardson, L. R. (1960). A longterm feeding study of irradiated chicken and green beans using the rat as experimental animal. Final Report, 1960.
153. Philips, A. W., Newcomb, H. R. and Shanklin, D. R. (1963). *Toxicol. Appl. Pharm.*, **5**, 273.
154. Thayer, D. W. (1984). *Cereal Foods Wld*, **29**(6), 353.
155. Kampelmacher, E. H. (1983). *Food Technol.*, June 1983, 117.
156. Gallica, C.-L., Paquin, J. and Sadat-Shafai, T. (1983). *Radiat. Phys. Chem.*, **22**(3), 759.
157. Bubl, E. C. and Butts, J. S. (1956). *Fed. Proc.*, **15**, 930.
158. Renner, H. W., Graf, U., Würgler, F. E., Altmann, H., Asquith, J. C. and Elias, P. S. (1982). *Food Chem. Tox.*, **20**, 667.
159. Read, M. S., Kraybill, H. F., Worth, W. S., Thompson II, S. W., Isaac, G. J. and Witt, N. F. (1961). *Toxicol. Appl. Pharm.*, **3**, 153.

Chapter 6

HIGH PRESSURE TECHNOLOGY AND MEAT QUALITY

J. J. MACFARLANE

CSIRO Division of Food Research, Meat Research Laboratory, Cannon Hill, Queensland, Australia

1. INTRODUCTION

Scientific study of the effects of pressure on biological systems appears to have commenced about a century ago following the discovery of organisms living in the depths of the sea at pressures of about 60 MPa.[1,2] The sensitivity of biological processes to pressure and the diverse response to pressure of these processes, some being accelerated and others retarded, were soon discovered. Such observations stimulated research into the underlying factors that control the response of these systems to pressure.[1-3]

Change of pressure affects only those reactions where there is a change of volume. These two parameters are related according to the equation[1]

$$k_p = k_0 \exp(-p\,\Delta V/RT) \qquad (1)$$

where k_0 is the rate at zero pressure, k_p the rate under pressure p, ΔV the volume of the activated complex minus the volume of the reactants, R the gas constant and T the absolute temperature. In the case of muscle contraction it is of passing interest to note that study of volume change appears to have preceded that of pressure effects by some centuries. Thus reference is made by Abbott and Baskin[4] to investigations carried out in the 17th century, in connection with the then current belief that muscle contraction involved an increase in volume because vital spirits moved from the nerve into the muscle and induced activation.

Although pressure and temperature are both important physical variables, investigations of the effects of temperature on biological systems far outnumber those of pressure. The relative neglect of pressure as a variable is no doubt largely due to the circumstance that outside the marine environment, most biological systems experience only small, usually inconsequential changes in atmospheric pressure.

Elevated pressures are employed in the food industry to enable heat processing operations to be carried out at temperatures higher than those attainable in an aqueous system at atmospheric pressure. Such 'pressure-cooking' applications are not of concern to the present discussion, which deals with the effects of much higher pressures of the order of 100 MPa (1000 atm). Application of high pressure to many foods, including meat, has long been known to delay putrefaction,[1] and more recently has been reported to initiate the germination of bacterial spores and then inactivate the germinated forms.[5] However the effects of high pressures on the structure of foods and exploration of possible beneficial effects of pressure treatments does not appear to have attracted attention until little more than a decade ago. Of all foods and food constituents, muscle and muscle proteins are probably the most responsive to pressure. This is because of the relatively high sensitivities to pressure of the glycolytic processes and of the associations between proteins of the myofibril.

It should be noted that in the pressure treatments under consideration here, samples are packaged in such a way that free air is excluded during pressure treatment. This avoids the complication of change of composition of samples due to dissolution of gases when pressure is increased. It also enables pressure to be changed relatively quickly because volume decreases only by about 3% on compressing water to 100 MPa. Typically fluid samples are pressure-treated when sealed in flexible plastic containers, or held in glass vessels individually sealed with a movable plastic plunger. Other samples are generally sealed in evacuated flexible plastic bags. A common method of applying pressure is to submerge samples in fluid in a pressure vessel then pump more fluid into the closed vessel. Thus non-fluid samples are not subjected to distortions and shears such as would occur if pressure was applied by, for example, squeezing between the plates of a mechanical press.

The SI unit for pressure is the pascal (Pa) and herein pressures are stated in megapascals (MPa). The relationships between the megapascal and other units commonly used for pressure are as follows:

$$\begin{aligned}1\,\text{MPa} &= 1\,\text{MN}\,\text{m}^{-2}\\ &= 10{\cdot}197\,\text{kg}\,\text{cm}^{-2}\\ &= 10{\cdot}000\,\text{bars}\\ &= 9{\cdot}869\,\text{atm}\\ &= 145{\cdot}038\,\text{lb}\,\text{in}^{-2}.\end{aligned}$$

2. EFFECTS ON THE STRUCTURAL PROTEINS OF MUSCLE

Little information has been published about the effects of pressure on the isolated structural proteins of muscle other than for the major proteins of the contractile system, namely actin, myosin and actomyosin. The findings for these latter proteins are outlined here because of the insight they provide into the mechanism of action of pressure treatment. Pressure exerts its effects on proteins through changes to the secondary, tertiary and quaternary structures of proteins, and not to any significant extent, at the pressures of interest here, through changes in the primary structure, i.e. rupture or formation of covalent bonds. (Of course where enzymes are activated by pressure, the associated reactions usually involve changes to primary structures.) The secondary and tertiary structures of proteins in an aqueous environment are stabilised by ionic and hydrogen bonds and by hydrophobic interactions. Breakage of ionic bonds results in a decrease in volume because of electrostriction effects of the free ions on water molecules. Therefore increase of pressure favours ionisation.[1,2] Hydrophobic interactions are disrupted by increase of pressure, whereas hydrogen bond formation is promoted; however, at pressures above 100–300 MPa these responses may be reversed.[6,7] An important consideration in the investigation of the effects of pressure on proteins in buffer solutions is the effect of pressure on the buffer itself. For example a phosphate buffer of pH 7 at atmospheric pressure, has a pH about 0·4 of a unit lower at 69MPa.[1,2] Wherever possible Tris-buffer should be considered because its pH is practically unaffected by pressure.[8]

2.1. Actin

Application of pressure has been shown to result in the depolymerisation of F-actin.[9–12] Ikkai and Ooi[11] found that at 25 °C F-actin starts to undergo irreversible denaturation at a pressure of approximately 147 MPa in the absence of adenosine triphosphate (ATP), and at approximately 245 MPa in its presence. However, if ethylene diaminetetraacetate (EDTA) is present with ATP, the pressure denaturation curve becomes almost the same as that when ATP is absent. They therefore suggested that in the presence of ATP pressure induces a reversible depolymerisation since G-actin is known to readily denature in the presence of EDTA, while F-actin does not. The pressure-induced depolymerisation of F-actin in the absence of ATP is favoured by decrease of temperature over the range 0–30 °C.[12] This is expected because the F → G transformation is associated with a negative enthalpy change.[13] Therefore the pressure of 147 MPa reported by Ikkai and Ooi for the initiation of depolymerisation of F-actin in the absence of ATP and at 25 °C, would be reduced substantially if a temperature of 0 °C was used. It also appears from a consideration of viscometric and electrophoretic evidence that, after the release of pressure, denatured G-actin monomers slowly reaggregate to form high molecular weight, non-filamentous polymers.[12]

2.2. Myosin

Observations have been made of the effects of pressure on myosin both during[14–18] and after pressure treatment.[10,19–22] Josephs and Harrington[14–16] have observed a shift in the myosin monomer–polymer equilibrium at the pressures developed in the ultracentrifuge. They carried out studies of myosin in a buffer of pH 8·3 and 0·18M KCl, under which conditions the polymeric species formed have a relatively sharp size distribution. A positive molar volume change of 380 ml mole^{-1} of monomer was estimated for the association process. Thus increase of pressure favours a shift in the equilibrium to the monomer form. They assumed that the positive volume change that occurs with the transition of monomer to polymer results predominantly from the release of compressed water about certain of the charged groups on the surface of the myosin molecules when these associate to form the ordered macrostructure of the polymer.[16] The equilibrium constant appears to be temperature independent and this supports the view that the association is brought about primarily by ionic bonding. Davis[17,18] has studied the effect of pressure on the thick filament–monomeric myosin equilibrium at pH 8·1. The myosin filaments studied appeared to be homogenous and had a central bare zone which is thought to arise from the anti-parallel configuration there of myosin molecules. He found that with increase of pressure the filaments shorten in concert towards the bare zone whilst maintaining the narrow length distribution seen at atmospheric pressure.

The effects of pressure on the myosin monomer–polymer equilibrium in the studies mentioned so far are reversible. However the pressures employed have been relatively low, ranging up to 35 MPa. At higher pressures effects on myosin that persist after the release of pressure are observed. Thus, although pressure treatment of myosin in the range 50·7–405·3 MPa does not affect the total number of SH groups of myosin, it increases the content of readily reacting SH groups.[19] Also pressure treatment of myosin in 0·6M KCl results in an increase in molecular weight.[10,20] This doubles after pressure treatment to 150 MPa for 1 h at 0 °C, and because there is no change in the radius of gyration, it has been suggested that overlapping dimers of myosin are formed.[20] The Ca^{2+} and K^{+}-EDTA ATP-ase activities of myosin are inhibited to about the same extent after pressure treatment[20] and this indicates that the pressure effect is not due to the selective destruction of one of the two reactive sulphydryl groups located at or near the active site of myosin. From studies of the lactoperoxidase-catalysed iodination of the tyrosyl residues of myosin before and after pressure treatment, it has been found that the accessibility of these hydrophobic groups is increased by pressure treatment at 150 MPa.[21] The tyrosyl residues of the heavy and the light chains are affected to a similar extent, indicating that pressure treatment has an effect over the entire surface of the myosin molecule. Pressure treatment at 75 MPa or higher for some minutes of myosin in 0·2–0·3M NaCl at about pH 6 increases the strength of the subsequently heat denatured samples.[22] Under these conditions of pH and ionic strength, unheated myosin at atmospheric pressure forms filaments. It was suggested that the alteration in heat-setting properties is due to depolymerisation of myosin filaments under pressure and a conformational change of the monomer so that it reaggregates in a different manner upon release of pressure.

2.3. Actomyosin

As might be expected in view of the effect of pressure treatment on the state of aggregation of actin and myosin, the aggregation of actomyosin is also affected by pressure. This is apparent from a decrease in turbidity of actomyosin solutions when under pressure,[23] a decrease in viscosity as a result of pressure treatment,[10,20] and changes in light scattering.[20] A matter of interest is whether, under pressure, actomyosin is dissociated into actin and myosin. Ikkai and Ooi[23] have observed that if it is, then the pressure effects on actomyosin should be the same as on the individual components and they carried out experiments to investigate whether this is so. They also studied heavy actomeromyosin and actosubfragment-1, which unlike actomyosin remain soluble in solutions of low ionic strength and neutral pH. From studies of pressure-induced changes in the adenosine triphosphatase activity of these compounds and their separate components, Ikkai and Ooi[23] suggest that in the absence of adenosine triphosphate heavy meromyosin protects actin from denaturation. Therefore dissociation of heavy actomeromyosin under pressure is thought unlikely. However, in the presence of adenosine triphosphate their results are consistent with a pressure-induced dissociation of actomyosin to actin plus myosin. That in the absence of adenosine triphosphate an association remains between the actin and myosin of actomyosin under pressure is also indicated by the results of studies of the lactoperoxidase-catalysed iodination of these proteins before and after pressure treatment at 150 MPa.[21]

3. TREATMENT OF PRE-RIGOR MUSCLE

3.1. Effect on Tenderness

If warm muscle taken from a freshly slaughtered ox or sheep is subjected to a pressure of the order of 100 MPa for a minute or more, then on removal from the pressure vessel the muscle will be found to be strongly contracted and very firm to the touch.[24,25] Data relating to pre-rigor pressure-treated muscle are presented in Table 1, along with that for muscle not pressure-treated. As a result of pressure treatment, the muscle shortened by about 35 % and this degree of shortening in non-pressure-treated muscle would be expected to result in considerable toughening of the cooked muscle.[26] However the shear force results indicate that pressure treatment improves the tenderness of muscle.[24,25,27,28] Although increased weep from muscle as a result of pre-rigor pressure treatment has been reported,[25] in other studies no or little change was detected.[27,29] Likewise the effect of pre-rigor pressure treatment on the water-holding capacity of muscle also is not clear because in one study it has been reported to be decreased,[25] whereas in another it appears to be possibly increased.[28] Cooking losses are lower from pressure-treated muscle than from that not pressure-treated.[24,25,28] Pre-rigor pressure treatment of muscle can result in an increase, over that of non-pressure-treated muscle, in the moisture content of the cooked meat.[24,28]

The reduced shear values of muscle resulting from pressure treatment are reflected in improved tenderness ratings by taste-panels.[24,27,28] For beef *Longissimus dorsi* muscle near maximum pressure-tenderising effect is

TABLE 1

RANGE OF VALUES FOR VARIOUS MEASUREMENTS ON PRESSURISED, CONTROL AND STANDARD[a] OX *Biceps femoris* MUSCLES FROM FOUR ANIMALS

Measurement	*Treatment*		
	Control	*Standard*	*Pressurised*[b]
Ultimate pH	5·61 ± 0·08	5·47 ± 0·04	5·49 ± 0·12
pH cooked meat	5·90 ± 0·10	5·69 ± 0·05	5·71 ± 0·23
Pressure-induced shortening (%)			35·0 ± 3·2
Total % shortening before cooking	24·5 ± 8·5		37·4 ± 4·9
Shortening due to cooking[c] (%)	14·9 ± 2·5	16·4 ± 1·0	7·5 ± 2·1
Weep (%)	1·3 ± 1·0[e]	1·0 ± 0·3[d]	2·0 ± 0·8[e]
Cooking loss (%)	35·3 ± 2·3	38·4 ± 3·0	31·1 ± 1·8
Moisture content cooked meat (%)	59·0 ± 2·2	57·3 ± 2·5	62·2 ± 1·5
Warner–Bratzler shear value cooked meat (kg)	13·57 ± 1·32	7·95 ± 1·82	4·11 ± 0·84

[a] Taken post-rigor from the side of the carcass that remains intact after removal from the carcass of one *Biceps femoris* muscle pre-rigor for control and pressure-treated samples.
[b] Conditions of pressurising: 103 MPa at 35 °C for 2 min.
[c] Measured on three samples only.
[d] After 1 day storage; muscles removed from carcass 1 day post-slaughter.
[e] After 2 days' storage.
Adapted from *Journal of Food Science* (1973), **38**(2), 294–8.

achieved by treatment at 103 MPa for 1 min (Table 2). Pressure treatment results in lower juiciness scores, even though it is likely that the moisture content of the cooked pressure-treated muscle is greater than that of non-pressure-treated muscle. From Table 2 it can be seen that pressure-treated muscle rates higher than non-pressure-treated muscle for acceptability; presumably the improvement in tenderness outweighs any adverse effects of loss of juiciness.

Many variables can affect the response of muscle to pressure, e.g. the actual pressure applied, the temperature during pressure treatment, the duration of pressure treatment, possibly the rate of pressure increase or decrease, restriction on the freedom of the muscle to contract, the extent of rigor development at the time of pressure treatment, the muscle treated, and the animal species from which the muscle was obtained. Treatment conditions to produce optimal meat quality, taking into account all the known relevant variables have not yet been defined. Improvement in the tenderness of meat by pre-rigor pressure treatment has been obtained over a wide range of treatment conditions, e.g. at pressures of 90 MPa up to the

TABLE 2

TASTE-PANEL SCORES FOR TENDERNESS, JUICINESS AND ACCEPTABILITY OF PRESSURISED AND NON-PRESSURISED (STANDARD[a]) BEEF *Longissimus dorsi* MUSCLES

Property	*Mean taste-panel score*[b]			
		Samples pressurised[c] *for*		
	Standard	*1 min*	*2 min*	*4 min*
Tenderness[d]	4·1 (0·22)	6·9 (0·27)	6·7 (0·15)	7·4 (0·22)
Juiciness[e]	6·0 (0·16)	5·6 (0·20)	4·5 (0·14)	5·2 (0·16)
Acceptability[f]	5·7 (0·21)	7·5 (0·24)	7·1 (0·12)	7·5 (0·21)

[a] Taken post-rigor from the carcasses, as indicated in footnote '*a*' of Table 1.
[b] Mean of results from five animals, standard error in brackets. The differences between standard and pressurised samples were all highly significant.
[c] Conditions of pressurising: 103 MPa, 35 °C, 1–4 min.
[d] 9 point scale: 1—very tough; 9—very tender.
[e] 9 point scale: 1—very dry; 9—very juicy.
[f] 9 point scale: 1—nauseating; 9—acceptable in preference to others.
Adapted from *Journal of Food Science* (1973), **38**(2), 294–8.

highest investigated, 150 MPa, and at temperatures of 25 °C up to the highest investigated, 40 °C.

3.2. Effects on muscle pH

A notable consequence of pressure treatment of warm pre-rigor muscle at 100–150 MPa for 1 min or more is a concomitant pH decrease of about 0·6 to 0·8 of a pH unit.[24,25] In rabbit *Longissimus dorsi* muscle the decrease can be about 1 pH unit.[30] Such large and rapid decreases in pH of the warm muscle might be expected to result in the meat exhibiting pale, soft and exudative characteristics,[31] and possibly the small increases in weep and decreases in water holding capacity that occur as a result of pre-rigor pressure treatment of muscle are manifestations of a tendency towards this condition. The extent of pH decrease is dependent upon the duration of pressure treatment, and the temperature of the muscle,[32] as shown in Fig. 1. It can be seen that the pH of sheep *Semimembranosus* muscle after 5 min pressure treatment at 30 °C is much lower than that of the control sample and continues to decrease rapidly during storage at atmospheric pressure to attain, in about 2 h, the ultimate pH value attained by the control after about 24 h. However, if pressure treatment is sustained for 3

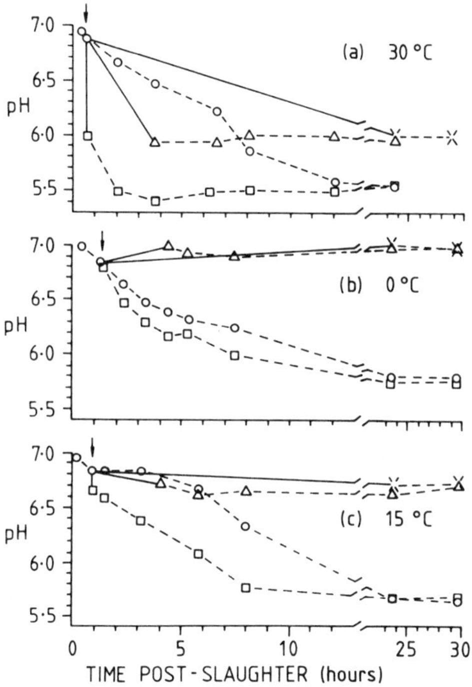

FIG. 1. Change in pH of sheep *Semimembranosus* muscle following pressure treatment at 150 MPa. Arrow indicates start of pressure treatment. (a) treatment and storage at 30 °C; (b) treatment and storage at 0 °C; (c) treatment and storage at 15 °C. Broken line, atmospheric pressure; full line, 150 MPa pressure. ○, non-pressure treated control sample; □, duration of pressure treatment, 5 min; △, duration of pressure treatment, 3 h; ×, duration of pressure treatment, 24 h (from Ref. 32).

or 24 h, muscle pH immediately after treatment is about the same as after treatment for 5 min, and remains constant during subsequent storage. Pressure treatment of muscle at 0 °C (Fig. 1(b)) for 5 min has perhaps only a slight accelerating effect on the rate of post-mortem pH decrease when compared with that for non-pressure-treated muscle. In contrast, if pressure treatment at 0 °C is maintained for 3 or 24 h, muscle pH remains nearly constant at its pre-pressurisation value during subsequent storage. No studies of the properties of meat with a high ultimate pH produced in this manner have been reported, but it is expected to exhibit the tenderness and high water-holding capacity usually associated with high pH meat. Even if a market for high pH meat exists, the pressure process does not appear to provide a commercially attractive method for its production. It could, however, provide a convenient method for producing relatively small quantities of high pH meat, as well as meat of intermediate pH (by delaying commencement of pressure treatment until muscle pH has decreased to a nominated value) as might be required for research studies. High pH meat can also be produced by 24 h pressure treatment of meat at 15 °C (Fig. 1(c)). At this temperature the effect of pressure treatment for 3 h on the extent of pH decrease during subsequent storage is unpredictable and it does not always decrease as shown in Fig. 1(c).[32]

3.3. Effect on Muscle Processes and Structure

The near ultimate pH values attained immediately after the pressure treatment at 100–150 MPa for about 2 min of warm (~30 °C) pre-rigor muscle indicate that post-mortem glycolysis has been greatly accelerated and the muscle is effectively in rigor. Accordingly, the ATP supply of the muscle is then also exhausted.[32] Studies of the effect of pressure treatment (150 MPa, 5–10 min, 35 °C) on the sarcoplasmic reticulum of beef, sheep and rabbit muscles have shown that the extra ATP-ase is largely destroyed by pressure treatment but the basal ATP-ase is scarcely affected.[30,33] Because the extra ATP-ase provides the energy for calcium uptake by the sarcoplasmic reticulum, its destruction in muscle would allow the concentration of Ca^{2+} in the sarcoplasm to increase and thus stimulate both contraction and glycolysis. It has been suggested that some of the Ca^{2+} needed for pressure-induced contraction may be released by mitochondria.[34] The key enzymes in the regulation of glycogen breakdown are phosphorylase, phosphorylase kinase and phosphorylase phosphatase, with phosphorylase being the rate-limiting enzyme. From a study of the effect of high pressure on these enzymes in rabbit muscles,[35] it appears that treatment at 35 °C enhances glycolysis through activation of phosphorylase, and this is synchronised with muscle contraction. The enhanced rate of glycogenolysis appears to involve loss of regulatory control of phosphorylase activity as a result of destruction of phosphorylase phosphatase.

Application of high pressure (150 MPa) to rabbit muscle at 0 °C results in a greater rate of loss of the activities of the three key enzymes regulating glycogen breakdown than treatment at 35 °C.[35] Because rabbit muscle at 0 °C does not contract with application of pressure to 150 MPa, the calcium level in the sarcoplasm then must be low and accordingly phosphorylase kinase will be inactive.[35] Thus there should be no increase in phosphorylase *a* levels and the pH of the muscle should not decrease. The relatively small changes in pH and phosphorylase *a* levels observed after pressure treatment at 0 °C for up to 10 min[35] are probably a consequence of the action, between the release of pressure and the taking of measurements, of enzyme activity that survives pressure treatment. Activation of these enzymes is consistent with the moderate contraction observed upon release of pressure after relatively brief treatments. Contraction indicates an increase in calcium level in the sarcoplasm. When sheep muscle is cooled to 0 °C it cold-shortens, but the contracted muscle at 0 °C can be stretched more readily when under pressure[32] and this indicates loss of the contraction stimulus. Using similar reasoning to that outlined above for rabbit muscle, the pH again should not decrease during pressure treatment. If the treatment is sufficiently prolonged key enzymes would be completely inactivated and no decrease of pH should then occur after pressure treatment, as has been found to be the case[32] (see Section 3.2).

Examination under the light microscope of muscle that has been subjected pre-rigor to pressure treatment reveals severe disruption of muscle structure with contraction band formation.[36] When examined using transmission and scanning electron microscopy changes in virtually all of the structural features of muscle are seen.[34] These include the separation of the sarcolemmal and endomysial sheaths, disruption of myofibrillar structure and increased interfibrillar and intermyofibrillar spaces, absence of glycogen granules, swollen and sometimes ruptured mitochondria, swollen sarcoplasmic reticulum and the absence of the H-zone and M-line characteristics of myofibrils. Some adjacent A-band filaments appear to have moved in opposite directions and this may have been facilitated by loss of M-line material.[37] Extensive disruption of endomysial sheath has been observed and it has been suggested that this would have a substantial weakening effect on the structure of muscle and could be one of the major reasons for the tenderising effect.[27]

Kennick and Elgasim[27] list the following four tentative explanations that have been made for the pressure tenderising effect: breakdown of myofibrillar structure; F–G transformations in actin or myosin or in both; creation of breaks in fibre structure as a result of massive contractions; early release of lysosomal enzymes. The first two effects are probably interrelated; evidence for this is provided by studies of pressure-induced length changes in strips of muscle under tension.[32] From Fig. 2 curves (i)–(iv) it is seen that pre-rigor muscle at 30 °C contracts immediately pressure is applied, but after 1–5 min under pressure the muscle lengthens again. This lengthening is not thought to be due to relaxation involving sliding of thick and thin filaments because the ATP supply of sheep *Semimembranosus* muscle is exhausted after pressure treatment for

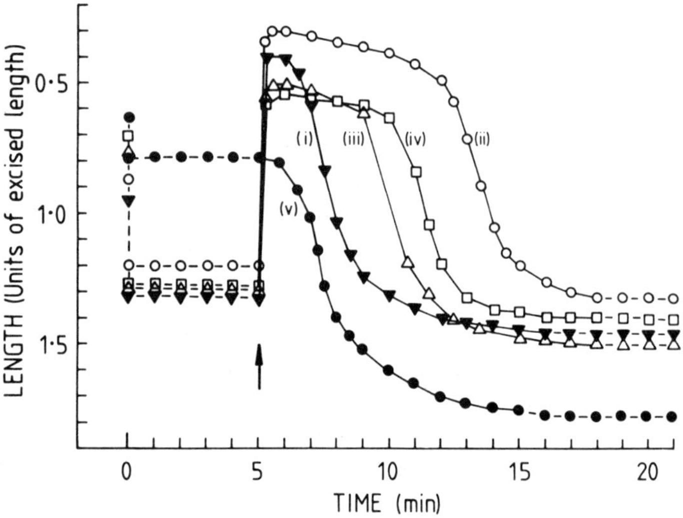

FIG. 2. Length changes in loaded sheep *Semimembranosus* muscle at 30 °C with application of pressure at various elapsed times *post mortem*. (i) 0·5 h *post mortem*, muscle pH 7·4, load 162 g cm^{-2}; (ii) 1·87 h *post mortem*, muscle pH 6·94, load 198 g cm^{-2}; (iii) 4·5 h *post mortem*, muscle pH 6·55, load 255 g cm^{-2}; (iv) 6·23 h *post mortem*, muscle pH 6·37, load 160 g cm^{-2}; (v) 26 h *post mortem*, muscle pH 5·63, load 132 g cm^{-2}. Fluid in windowed pressure vessel, water. Broken line, atmospheric pressure; full line, 150 MPa pressure (arrow indicates initial application of pressure) (from Ref. 32).

2 min.[32] The lengthening, therefore, presumably occurs because of disruption and weakening of myofilaments under pressure. In the case of thin filaments a decrease in their stability under pressure when the ATP supply is depleted would be expected because of the increased sensitivity to pressure of the F → G transformation when ATP is not present[11] as referred to in Section 2. If myofilaments are weakened *via* such a mechanism, then application of pressure to post-rigor muscle should also reduce its tensile strength and this has been found to occur (Fig. 2—curve (v)). The resulting disruption of the weakened myofibrils during contraction may account for the lower shear values obtained for pressure-treated pre-rigor muscle than would be expected from muscle contracted to the same extent by cold-shortening.[38]

3.4. Comments Relating to the Practicability of the Treatment

High pressure treatment of pre-rigor muscle in principle provides a way to achieve two desirable objectives, namely the rapid processing of hot boned meat with its potential for energy and cost savings, and the promotion of meat tenderness. The main operations likely to be needed for a pressure tenderisation process are:

(a) hot-boning of the carcass, or those portions of the carcass to be pressure-treated, promptly after the completion of dressing operations;
(b) vacuum-packaging of the primal cuts to be treated;
(c) pressure treatment of the warm, pre-rigor primals;
(d) chilling or freezing of the treated meat.

Operations (a), (b) and (d) have been studied in connection with the present or proposed use of hot-boning technology in the industry.[39,40] Operation (c) is the new issue to be considered with respect to the practicability of commercial pressure tenderisation. However, should the process appear commercially viable, further considerations of the other operations could be required to develop a properly integrated procedure.

The feasibility of operation (c) above is under study at the Oregon State University.[41] There the treatment has been investigated initially using a vessel of about 2 kg capacity, and subsequently with one approximately 30 cm in diameter and 61 cm long.[27,29] The latter vessel is capable of holding several large boneless cuts representative of those produced in commercial practice. It has been found that pressure treatment remains effective when used on the larger cuts, and that purge (weep) losses are about the same as for non-pressure-treated cuts. Whole sides of beef have been hot boned to give twelve major wholesale cuts that were vacuum packaged and pre-rigor pressure treated.[29] No significant differences were found in total trimmed retail yields betwen sides processed using high pressure treatment and those processed using conventional methods. An altered shape of cuts as a consequence of pressure treatment in some cases reduces the yield of steaks. However, it has been suggested that this could be overcome by developing different packaging techniques for cuts before they are pressure-treated, as well as changing cutting techniques.[27,29]

Investigators at Oregon have studied the effect of pressure treatment on some attributes of meat which are relevant to commercialisation of the process. Pre-rigor pressure-treated meat, stored at 1–3 °C under simulated retail conditions, has been found to have a better appearance for the first four days of storage than muscle from carcasses that were conventionally processed.[42] With regard to protein functionality it appears that when assessed before onset of rigor mortis in control samples, the solubility and emulsification capacity of proteins in pressure treated samples are lower than they are in controls.[43] However, at 24 h and 168 h *post mortem*, no differences in emulsifying capacity between pressure-treated and control samples are evident. Studies of the effect of pre-rigor pressure treatment of beef muscles on protein quality have shown that the treatment does not affect the apparent biological value or the net protein utilisation, but significantly improves protein digestibility.[44] No effect of the treatment on total essential and nonessential amino acid contents of meat was detected.

4. EFFECT ON POST-RIGOR MUSCLE

Although pressure treatment of warm pre-rigor meat is effective for avoiding myofibrillar toughness in meat, a treatment that is effective when applied to post-rigor muscle obviously would be of greater potential usefulness. There are at least two ways by which application of high pressure might improve the tenderness of post-rigor muscle. One is by affecting enzyme systems[45] responsible for the degradation of the structural proteins of muscle. Enzymatic processes involve volume changes which can be either positive or negative[46] and hence will be retarded or accelerated by pressure treatment accordingly. Also, a calcium activated protease[47] has been implicated in meat tenderisation and it seems plausible that Ca^{2+} might be freed by pressure treatment and activate this enzyme. Another way is by causing irreversible physical change in the myofibrillar proteins, in particular disaggregation of actin and myosin as referred to in Section 2. Accordingly, studies have been carried out on the effect of pressure treatment on post-rigor muscle to explore these possibilities.

4.1. Pressure Treatment at 0°–20°C

If meat is vacuum-packed and then pressure treated at up to 150 MPa for some minutes, there is little or no change in its appearance. With increase in the duration of pressure treatment to some hours, its appearance becomes brownish, presumably because of denaturation of proteins. Pressure treatment has only a small lasting effect on the pH of the meat, treatment at 150 MPa for 3 h at 0 °C causing an increase of little more than 0·1 pH unit.[48] Such a pressure treatment results in a small increase in Warner–Bratzler shear force values only in contracted muscle that is subsequently no more than lightly cooked, e.g. at 50 °C.[48] The increase is thought to be due to the occurrence of a pressure-induced denaturation and coalescence of the myosin of thick filaments. The effect is no longer apparent in well cooked (80 °C) muscle presumably because of the then much larger thermally-induced increases in shear values in both pressure-treated and non-pressure-treated samples.

Although the shear results give no indication of a substantial effect of pressure treatment on the structural integrity of muscle, studies of ultrastructure show otherwise.[37,48] These have revealed that pressure treatment results in an apparent loss of integrity and disaggregation of I-band filaments (Fig. 3(b)). Evidence of translocation of thin filament components under pressure is provided by the lack of a defined H-zone. As this zone is the region where there is no overlap of thick and thin filaments, this result indicates that under pressure depolymerised actin can move along the length of the A-band and, when pressure is released, associates with the myosin of thick filaments. Therefore, it appears that pressure treatment has a disaggregating effect on the actin of thin filaments similar to that it has on isolated F-actin referred to in Section 2. Presumably other constituents of the I-band, notably tropomyosin and troponin, are also disaggregated by pressure treatment. As mentioned in Section 2.3, in the

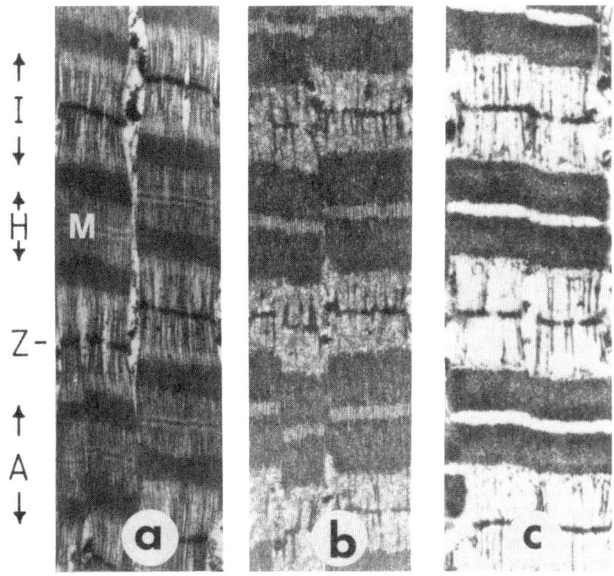

FIG. 3. Ultrastructural appearance of longitudinal sections of sheep *Semimembranosus* muscle. (a) control sample; (b) pressure-treated at 150 MPa for 18 h at 0 °C; (c) heated at 60 °C for 30 min under a pressure of 150 MPa. Z, Z-line; I, I-band; H, H-zone; A, A-band; M, M-line.

absence of ATP, actin when associated with myosin as in actomyosin appears to be stabilised to pressure. If the actin bound to myosin by rigor bonds in the region of overlap of thick and thin filaments is also stabilised, then this portion of the actin would not be freed and able to migrate under pressure. Disaggregation of I-band proteins could account for the decrease in the tensile strength of post-rigor muscle that occurs soon after application of pressure (Fig. 2 curve (v)).[32] Another change to the myofibrillar structure as a result of pressure treatment is removal of M-line protein,[37] revealing a clearly defined pseudo-H band (Fig. 3(b)). Therefore some or all of the proteins of this region, such as creatine kinase and myomesin[49] are disaggregated under pressure.

Further evidence of major change to the actin in muscle as a result of pressure treatment has been obtained by means of differential scanning calorimetry.[48] This technique provides information about muscle proteins by measuring thermal transitions that occur in the intact muscle when it is heated. Such investigations reveal that prolonged pressure treatment of muscle at about 150 MPa results in the elimination of the thermal transition attributable to actin.

The disruption of the I-band by pressure treatment without a concomitant decrease in Warner–Bratzler shear values of the cooked meat appears to support one of the theories that have been advanced to account for the rising phase of the contraction state versus toughness relationship in terms of the thick and the thin filaments.[48] The theory supported is that in which the rising phase of toughness is related to an increasing incidence of sarcomeres 1·5 μm or less in length, in which thick filaments of myosin are compressed into the Z-line thus eliminating the I-band as a zone of weakness.[50] Recently it has been proposed that contraction state toughness is due to another component of the myofibril, termed 'gap filaments' and that thick and thin filaments do not contribute because they are disaggregated when meat is cooked. This theory is referred to again in Section 4.2.

The studies, referred to in Section 2, on isolated muscle proteins show that myosin as well as actin is disaggregated by application of a pressure of 150 MPa that was often used in the above studies of muscle. However, the ultrastructural appearance of muscle after prolonged pressure treatment (18 h, 150 MPa) indicates that the thick filaments are still intact (Fig. 3(b)). Therefore, if these filaments are disaggregated under pressure, the constituent myosin molecules are unable to move sufficiently far from their original location to interfere with the reforming of filaments upon release of pressure. Alternatively, in muscle the thick filaments may be stabilised to pressure disaggregation by their environment (pH, ionic strength) or by other components.

4.2. Pressure Treatment at 45–60 °C

In contrast to the lack of a tenderising effect of pressure treatment at 0–20 °C, when meat is heated to 45–60 °C under pressure a marked improvement in tenderness is obtained[51] (Fig. 4). Prolonged heating of

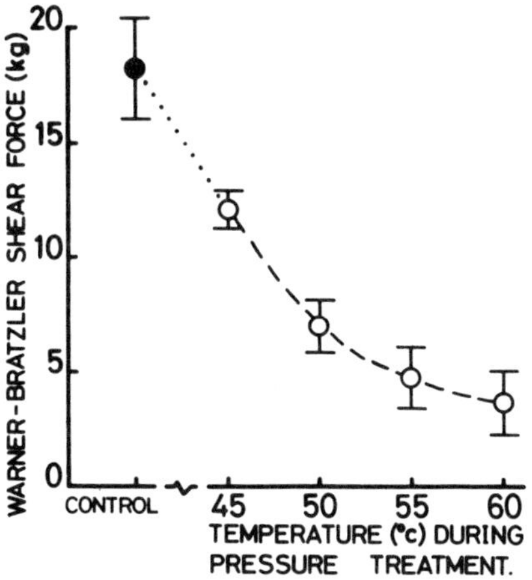

FIG. 4. Warner–Bratzler shear force values obtained for cold shortened beef *Semitendinosus* muscles pressurised for 1 h at 150 MPa with the pressure vessel at 45, 50, 55 or 60 °C. All samples were at 25 ± 1 °C before pressurising. Samples were cooked, after pressure treatment, at 80 °C for 1·5 h. Control samples were neither pressurised nor heated before cooking. Vertical lines through points, ± standard deviation. Reprinted from *Journal of Food Science* (1977), **42**(1), 132–5. Copyright © by Institute of Food Technologists.

meat prior to pressure treatment, for example for 2 h or more at 55 °C, or 1 h or more at 60 °C, causes the treatment to lose some of its effectiveness in reducing shear force values. To account for these observations in terms of a pressure-induced dissociation of the native proteins a reaction scheme, similar to the following, has been proposed:[51]

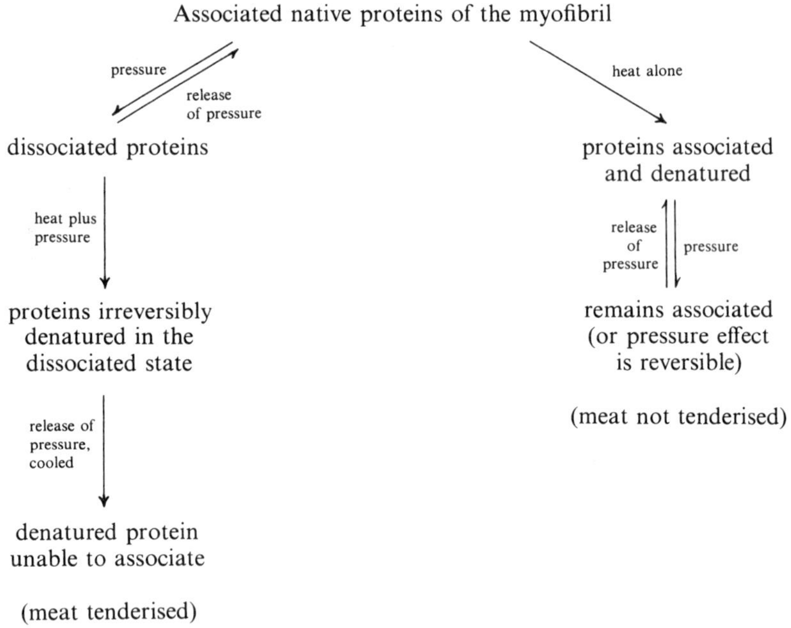

FIG. 5. Scheme involving a pressure-induced dissociation of proteins to account for the tenderisation of meat by a combined pressure and heat treatment.

However, a mechanism in which endogenous enzymes are activated by the application of pressure at temperatures up to about 60 °C, near which temperature they are slowly thermally inactivated is also possible. The combined use of pressure and heat to tenderise meat is subsequently referred to as P–H treatment.

The P–H tenderisation process can be hastened, possibly with some risk of a small loss in effectiveness, by heating the meat to, say, 55 °C before loading it into the pressure vessel, also at 55 °C, then applying pressure for about 2 min.[51] The effectiveness of various combinations of meat temperature and duration of pressure treatment in reducing shear force values has not been thoroughly investigated, and it is possible that near maximal tenderisation may be achievable using briefer pressure treatments than the 2 min referred to above.

A feature of the P–H treatment is its effectiveness for overcoming toughness associated with cold-shortened muscle. Tenderising by ageing is not effective against this form of toughness[52] and the P–H process is the only method known to be effective on intact post-rigor muscle. After P–H treatment muscle samples that have entered rigor in widely different contraction states show similar shear values.[51] This finding does not support the suggestion[53] that contraction toughness is a consequence of changes in the arrangement of collagen fibres. A comparison of tenderisation by pre-rigor pressure treatment with that by post-rigor P–H treatment,[36] has shown both treatments to be equally effective, which is indicative that both destroy the same structural components of muscle. From an examination of the ultrastructure of contracted muscle it appears that P–H treatment destroys the integrity of A-bands, as evidenced by gaps that can be seen at the middle of the sarcomere at the region of the M-line.[36] Such gaps can be well defined in P–H treated, stretched muscle (Fig. 3(c)).

Objective assessments of the effect of P–H treatment on the mechanical properties of meat using compression, adhesion and Warner–Bratzler shear measurements, indicate that the myofibrillar component of toughness is destroyed but that of the connective tissue is not affected.[54,55] Subjective assessments of the effect of P–H treatment on a range of muscles show that tenderness is limited by connective tissue toughness.[54] P–H treatment also decreases the juiciness of meat.

P–H Treatment of Connective Tissue

To further substantiate the apparent lack of effect of the P–H treatment on the strength of intramuscular connective tissue, studies have been carried out on isolated connective tissue sheath.[56] The material that has been investigated is sheep epimysial/tendinous tissue, subsequently referred to here as tendon, obtained from the exterior surface of the *Longissimus dorsi* muscle. In these studies, at the completion of the P–H treatment, pressure was maintained until the sample cooled to $<40\,°C$.

By carrying out the P–H treatment over a range of temperatures, it has been found that under a pressure of 150 MPa, onset of the heat shrinkage reaction in tendon occurs at about 64 °C compared to 58 °C at atmospheric pressure.[56] Therefore, it appears that application of pressure stabilises tendon to heat denaturation. Measurements of stress-relaxation that occurs when the temperature of tendon under tension is raised, and of the enthalpy changes associated with the thermal transition that occurs in tendon at about 60 °C, also show that increase of pressure stabilises tendon to heat denaturation.[56] This is in accord with the known stabilisation of the helical structure in native collagen by hydrogen bonds,[57] and with an expected volume increase when these are broken during the collagen shrink reaction.

P–H Treatment and Gap Filaments

In the above discussion the myofibrillar component of meat toughness has been related to the behaviour of the thick and the thin filaments on cooking, the thick filaments being considered to strengthen and the thin filaments to disintegrate. However, recently, Locker[58] has claimed that the thick filaments are dissociated when meat is cooked, the myosin then forming an actomyosin gel with the actin from disintegrated thin filaments. He therefore considers that these filaments cannot be responsible for the development of myofibrillar toughness and proposed that it is due to another component of the myofibril, the gap filament. According to this theory, the P–H treatment achieves its effect by disruption of gap filaments and not thick filaments. Locker has investigated the effect of a 60 MPa P–H treatment on gap filaments and found that in cooked meat gap filaments survive morphologically, but are more extensible and break at lower loads than in control, heat only samples.[58] The gap filaments are thought to be weakened in aged meat as a result of attack by endogenous proteases. The rate of meat tenderisation by ageing is maximal at 60 °C,[59] and because this temperature is that preferred for the P–H treatment, it appears possible that during P–H treatment, pressure accelerates proteolysis of gap filaments.

However, the claim that thick filaments are dissociated when meat is heated is not supported by the results of measurements of the tensile strength of heat-coagulated extruded myosin fibres.[60] These measurements reveal that the tensile strengths exhibited by these fibres match the breaking strengths reported for cooked meat. The orientation of myosin molecules in the extruded fibres no doubt is very different from the ordered structure present in the myofibril, but the result indicates that when myosin is heat denatured under conditions that favour fibre formation, generalised dissociation of myosin molecules does not occur. Connectin, or titin, is thought to be the major protein constituent of gap filaments, and a recent report of the fate of connectin in muscle under various conditions also raises difficulties for the gap filament theory of toughness.[61] For example connectin appears to be broken down in muscle cooked at 80 °C but the muscle retains its toughness. Also connectin appears to be equally vulnerable to the action of heat in meat samples that exhibit large differences in toughness due to differences in muscle contraction state.

4.3. Applications of P–H Treatment

To date serious consideration does not appear to have been given to possible commercial applications of the treatment. However P–H treatment has been found to be useful in investigating the contributions to meat toughness of the myofibrils and the connective tissues. Its usefulness arises from the virtually complete destruction of the myofibrillar component of toughness by P–H treatment regardless of the contraction state of the muscle. Therefore, in principle, from objective or subjective assessments of toughness before and after P–H treatment, the contributions of both components can be estimated. Examples of ways in which the P–H treatment might be used follow.

Tenderisation by Ageing

Although ageing, or conditioning, is one of the most widely employed methods for improving the tenderness of meat, it is difficult to assess whether the full potential tenderising effect of ageing is being realised. The usefulness of the P–H treatment for determining the effectiveness of ageing for destroying the myofibrillar component of toughness has been

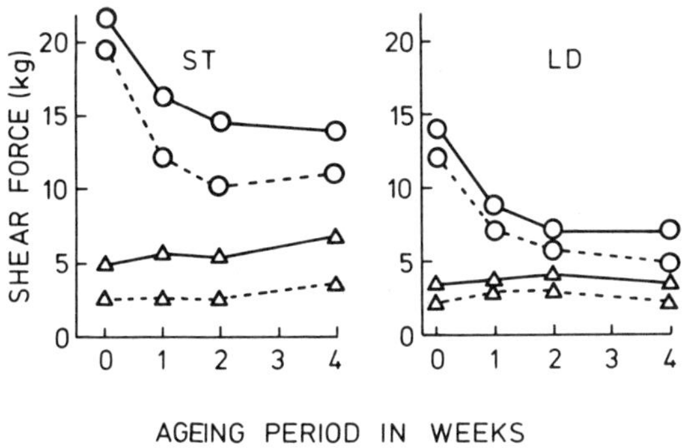

FIG. 6. Effect on peak force values (unbroken lines) and initial force values (broken lines) of applying a pressure-heat treatment to beef *Semitendinosus* muscle samples (ST) and beef *Longissimus dorsi* muscle samples (LD). The ST muscles were removed from the carcass pre-rigor and cold-shortened; the LD muscles were removed from the carcass post-rigor. △, pressure-heat treated samples; ○, non-pressure-heat treated (control) samples. Least significant differences (5% level):

	Ageing	Pressure-heat
ST muscle, peak force	1·8	1·7
initial yield force	2·8	2·2
LD muscle, peak force	0·8	0·7
initial yield force	0·8	0·8

Reprinted from *Journal of Food Science* (1980), **45**(2), 276–8.

demonstrated in experiments in which the P–H treatment was applied to fresh and aged meat.[62] Prior ageing of muscle did not result in a further decrease in the peak force values of P–H treated samples (Fig. 6). This indicates that ageing does not affect structural components of muscle other than those affected by the P–H treatment. In comparison to P–H treated muscle, it can be seen that control cold-shortened *Semitendinosus* muscle retains considerable myofibrillar toughness after 4 weeks' ageing (Fig. 6), as expected in view of the known resistance of contracted muscle to tenderisation by ageing.[52]

It has been observed that the P–H treatment becomes slightly less effective after near-extreme ageing treatments (4 weeks, 5–6 °C).[62] Neither of the two possibilities advanced for the lessened effectiveness, namely an increase in the strength of connective tissue with ageing, and a decrease in the effectiveness of the P–H treatment on the myofibrils, can be completely discounted. On the evidence available the latter is thought the more likely explanation. The P–H treatment has been shown to be effective on, and useful for investigating ageing effects in, meat from electrically stimulated carcasses.[63]

Influence of Muscle pH

The primary effect of pH on meat tenderness is believed to be via the myofibrillar structure of muscle because of the large effect of pH on the water-holding capacity of the myofibrillar proteins. However, it is possible that pH might exert a considerable effect on the response of the connective tissues to heat, and this aspect has been investigated using sheep muscles with pH's ranging from 5·5 to 7·5 approximately.[64] The shear force values of the non-P–H treated muscles decrease with increase of the pH of the muscle from 5·5 to 7·0. After P–H treatment, shear force values are almost unaffected by muscle pH, and are similar in value to those for muscle of pH 7. It therefore appears that change of muscle pH has little, if any, effect on the shear force attributable to the connective tissue of muscle, and that a myofibrillar component of toughness is not evident in muscle of pH 7.

Animal Comparison

The myofibrillar component of toughness is sometimes referred to as processing toughness because events post-slaughter determine its magnitude. On the other hand the connective tissue component is sometimes referred to as animal toughness because its magnitude is largely inherent in the animal when presented for slaughter. With advances in the technology for avoiding processing toughness, animal toughness becomes of greater concern. An area of interest is the differences that may exist between breeds or species. In a comparison of beef and buffalo meat it has been found that P–H treatment enables differences between these species in the contribution of connective tissue to toughness to be more readily detected.[65] Similarly, P–H treatment should be useful for investigating animal toughness differences arising from other factors, animal age in particular.

5. EFFECTS ON MANUFACTURING PROPERTIES

In manufactured meat products, muscle proteins are used to achieve various desired effects such as stabilisation of dispersed fat, retention of water, and binding together of the ingredients of the product. The effectiveness of these proteins for achieving the above effects is improved by the presence in the product of salts such as sodium chloride and sodium pyrophosphate. It is thought that the improvement with salt is primarily due to swelling of the myofibrillar proteins or their solubilisation, depending on the conditions employed.[66–68] Changes in electrostatic interactions are undoubtedly involved and therefore sensitivity of the meat–salt system to pressure would be expected. That this is so has been shown in studies of the effects on protein solubility (estimated as 'expressed soluble protein', ESP), and water-holding capacity (WHC) of pressure treatment of homogenates of meat in saline solution[69] (Fig. 7). It appears that muscle can be broadly divided into two categories according to whether pressure has a marked or only a slight effect on WHC, and that even when the non-pressure-treated muscles are compared, those in the former category generally have higher WHC. Increase in WHC as a result of pressure treatment may result in an apparent decrease in ESP (Fig. 7) presumably as a result of a decreased amount of free fluid. However, after treatment at 150 MPa, the ESP from all samples is increased. In homogenates consisting of 1 part meat and 10 parts 0·5M (approximately) NaCl, WHC has been increased after pressure treatment for 1 min at 25 MPa,[69] the lowest pressure above atmospheric pressure that has been studied. The increase in ESP is accounted for by an increase in myofibrillar proteins solubilised.[69,70] The occurrence of the pressure effect is pH dependent, being negligible below pH 5·3 and almost so above pH 7·0 where the higher muscle pH itself results in much higher ESP values.[69] Pressure-induced solubilisation of isolated myofibrils in 0·5M KCl is also pH dependent,[70] in this case showing a marked increase at pH 6 and higher values until pH 10–11 where the solubility of the non-pressure-treated

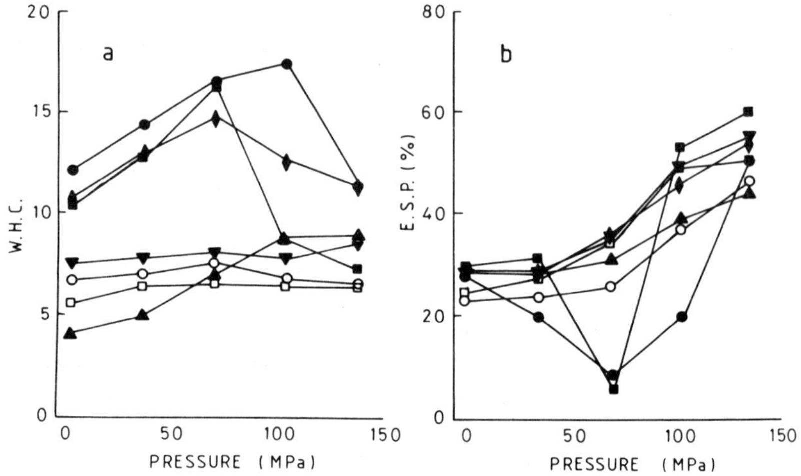

FIG. 7. Changes following pressurisation in water holding capacity, WHC (a), and expressible soluble protein, ESP (b), of homogenates of meat (1 part) in sodium chloride solution (3 parts). Salt concentration in aqueous phase, 0·5M. Samples pressurised at 0 °C for 5 min. Samples from different animals are indicated by different symbols.

Reprinted from *Journal of Food Science* (1974), **39**(3), 542–7.

myofibrils is equally high. Because imidazole groups are titrated around pH 6 it appears that this group is involved in the pressure effect.[69,70]

Pressure treatment at up to 150 MPa also has been shown to increase the binding achieved between meat particles in patties when they are subsequently cooked.[71] Binding strength has been estimated from measurements of the tensile strength of samples cut from patties. Studies have been carried out using finely comminuted meat and 10 % added water, with or without added NaCl. The magnitude of the increase in binding strength depends on the intensity and duration of pressure treatment, the concentration of NaCl in the patty and the pH of the patty. Interestingly, binding strength of patties can be increased by application of 150 MPa at all four added NaCl concentrations investigated, namely 0, 0·5, 1 and 3 %. The proteins thought to be the most effective for binding, myosin and actomyosin, are not expected to be solubilised after pressure treatment at low ionic strengths, e.g. in the 0 and 0·5 % added NaCl samples. The increase in binding strength of patties under these latter conditions can be accounted for by the occurrence of a pressure-induced disaggregation and conformational change of proteins, so that upon release of pressure a structure different from that in non-pressure-treated patties is obtained. Such a mechanism has been proposed to account for an alteration in the heat-setting properties of myosin filaments as a result of pressure treatment.[22]

Although pressure treatment results in an increase of binding strength of patties at all the added NaCl levels investigated, cooking losses increase at 0 and 0·5%, but decrease at 1 and 3%, added NaCl. The decreased cooking losses in the latter samples are in accord with protein swelling and solubilisation leading to increased binding strength. The increased cooking losses at zero or low added NaCl may be related to an altered conformation of proteins as referred to in the discussion of tensile strength changes.

Polyphosphates are often added to meat products along with sodium chloride to improve WHC and binding properties. As these properties in meat patties also appear to be improved by pressure treatment, a comparison of the two procedures is of interest. Another question is whether pressure treatment of patties containing NaCl and polyphosphate confers additional benefits. To date the only products studied have contained 1% added NaCl with or without 0·5% pyrophosphate, so whether the findings have general applicability is not known. Pressure treatment increases the binding strength of patties at lower pH values more than pyrophosphate does, but pyrophosphate remains effective at higher pH values (Fig. 8).[71] Pressure treatment markedly increases the binding strength in patties with pH values of 5–6·5 approximately and containing NaCl and pyrophosphate. Pressure-treated patties with NaCl but not pyrophosphate show lower cooking losses over the pH range 4·7–6·2, but in those with added pyrophosphate, cooking losses are not decreased until their pH is about 5·5 or higher.

Higher WHC's and perhaps binding strengths of patties might be obtained by using pyrophosphate in combination with more than 2% NaCl, to take advantage of the synergistic action of these two ingredients.[72] However, such NaCl concentrations may impart an undesirably strong salty taste to the product. They may also be objected to because sodium ion has been implicated in the development of hypertension.[73] Accordingly the pressure treatment could be of interest to maintain or improve protein functionality where it is desired to reduce the sodium ion content of processed meats.

Hamm[67] mentions hypotheses that have been advanced to explain the effect of phosphates on the WHC of meat and it is of interest to consider whether the mechanisms that have been suggested could account for the

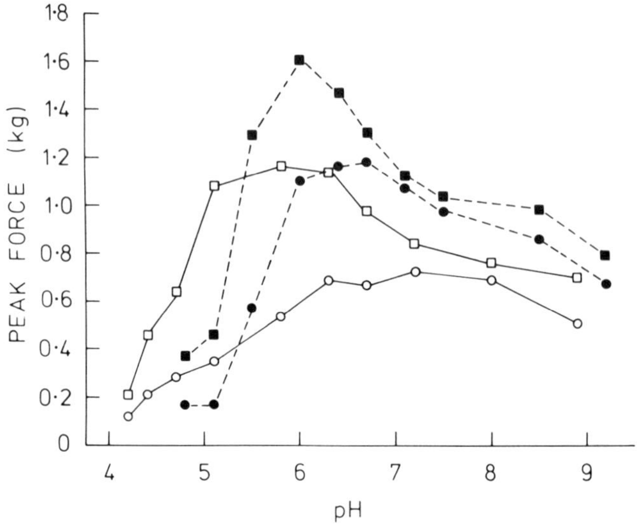

FIG. 8. Comparison of the binding strengths, as measured by tensile peak force values, of pressure-treated and non-pressure-treated beef patties of various pH values, with and without added tetrasodium pyrophosphate (TSPP). All patties were cooked at 70 °C for 30 min. Circles, non-pressure-treated samples; squares, pressure-treated at 150 MPa for 20 min; open symbols, no added TSPP; solid symbols, with 0·5 % TSPP. Salt concentration in the aqueous phase, 1 %. Least significant difference ($p = 0{\cdot}05$), 0·1 kg at the same pH and TSPP level; 0·15 kg otherwise (from Ref. 71).

action of pressure treatment. The relevant mechanisms are those that are suggested to operate in the presence of NaCl. One proposal for the action of phosphates is for an increased solubility of muscle proteins caused by the dissociation of actomyosin. Although pressure treatment appears to promote the solubilisation of actomyosin, as indicated in Section 2.3 dissociation into actin and myosin seems doubtful. A second proposal, that the decisive factor is the relatively high ionic strength brought about by the addition of polyphosphates, may be likened to a pressure effect on a Donnan equilibrium that may exist between the muscle proteins and the salt solution. However, in the absence of other changes, the effect of pressure on a Donnan equilibrium would be reversible and therefore would not persist after the release of pressure. A third proposal involves breaking of divalent cation (e.g. Ca^{2+}, Mg^{2+})–protein bonds. Breakage of such bonds under pressure would appear likely because of electrostriction effects of the ions on surrounding water molecules. The binding sites would then be able to move apart, and if both bonds of the bridging bivalent ion are broken, this ion can also diffuse away. Upon release of pressure, the probability of divalent ion bridges reforming would be reduced because of changes in the conformation of proteins due to the added NaCl. As a result, increases in WHC and solubility of protein could persist after the release of pressure.

6. CONCLUDING REMARKS

The application of pressures up to 150 MPa to muscle and related systems has revealed responses of considerable magnitude and potential usefulness. Thus post-mortem glycolysis can be controlled to the extent that at one extreme it can be greatly accelerated so that it is virtually complete within a few minutes, and at the other it can be effectively inhibited from the time that the meat has been cooled to about 15 °C or lower and pressure applied. Therefore meat of virtually any desired pH value in the usual post-mortem range can be produced. However pressure treatment itself will cause changes, which at present are ill-defined, in other structures of muscle and this needs to be borne in mind if pH is controlled by this method.

If conditions of pressure treatment favourable for accelerated post-mortem glycolysis are employed, then virtually in-rigor meat with no myofibrillar toughness, can in principle be rapidly processed from hot-boned pre-rigor meat obtained from carcasses immediately after the completion of dressing operations. Some lessening of connective tissue toughness in, and cooking losses from, muscle may also be achieved, but these as well as other characteristics of the muscle require further investigation. Should commercial application of the process be considered, an obvious requirement is a hot-boning procedure for carcasses.

A procedure also has been developed for the pressure-tenderisation of post-rigor muscle. It provides the only means by which toughness due to muscle contraction can be eliminated in intact post-rigor muscle. A disadvantage of the process is the necessity to heat the meat to about 60 °C during pressure treatment. The treated meat has a cooked appearance which would limit outlets for it even if the process was practicable in other respects.

Another aspect of the effect of pressure on muscle proteins is illustrated by the studies of pressure effects on homogenates of meat in saline solution. Pressure influences the action of salt in the swelling of muscle tissues and the dissolution of muscle proteins. Pressure treatment can also improve the cohesion between meat particles in reformed meat type products even at low ionic strengths.

Meat scientists and technologists have largely either overlooked, or discounted the importance of, pressure as an experimental variable. Perhaps it was thought that worthwhile effects could not be obtained, or even if they could, high pressure processing appeared impracticable. However, the potential usefulness of the effects described above, and the possibility of improving them or finding others, appears to justify further study of the influences of this variable. Even in the absence of commercial application of pressure processing, pressure treatment is a useful investigational tool for the meat scientist.

7. ACKNOWLEDGEMENTS

I am grateful to Dr D. J. Horgan and Dr R. P. Newbold for helpful comments in the preparation of this chapter.

8. REFERENCES

1. Johnson, F. H., Eyring, H. and Polissar, M. J. (1954). *The kinetic basis of molecular biology*, John Wiley and Sons, New York.
2. Johnson, F. H. and Eyring, H. (1970). In: *High pressure effects on cellular processes* (Zimmerman, A. M. (Ed.)), Academic Press, New York, p. 1.
3. Kitching, J. A. (1972). In: *The effect of pressure on organisms* (Sleigh, M. A. and Macdonald, A. G. (Eds)), University Press, Cambridge, p. 473.
4. Abbott, B. C. and Baskin, R. J. (1962). *J. Physiol.*, **161**, 379.
5. Clouston, J. G. and Wills, P. A. (1969). *J. Bact.*, **97**, 684.
6. Suzuki, K. and Taniguchi, Y. (1972). In: *The effect of pressure on organisms* (Sleigh, M. A. and Macdonald, A. G. (Eds)), University Press, Cambridge, p. 103.
7. Suzuki, K., Miyosawa, Y., Tsuchiya, M. and Taniguchi, Y. (1968). *Rev. Phys. Chem. Japan*, **38**, 63.
8. Distèche, A. (1972). In: *The effect of pressure on organisms* (Sleigh, M. A. and Macdonald, A. G. (Eds)), University Press, Cambridge, p. 27.
9. Estes, J. E. (1974). *Fed. Proc. Fed. Am. Soc. Exp. Biol.*, **33**, 1522.
10. Ivanov, I. I., Bert, Yu. N. and Lebedeva, N. A. (1960). *Biokhimiya*, **25**, 505.
11. Ikkai, T. and Ooi, T. (1966). *Biochemistry*, **5**, 1551.
12. Macfarlane, J. J. and McKenzie, I. J. (1983). In: *Actin, structure and function in muscle and non-muscle cells* (dos Remedios, C. G. and Barden, J. A. (Eds)), Academic Press, Sydney, p. 129.
13. Asakura, S., Kasai, M. and Oosawa, F. (1960). *J. Polym. Sci.*, **44**, 35.
14. Josephs, R. and Harrington, W. F. (1966). *Biochemistry*, **5**, 3474.

15. JOSEPHS, R. and HARRINGTON, W. F. (1967). *Proc. Natl. Acad. Sci. (USA)*, **58**, 1587.
16. JOSEPHS, R. and HARRINGTON, W. F. (1968). *Biochemistry*, **7**, 2834.
17. DAVIS, J. S. (1981). *Biochem. J.*, **197**, 301.
18. DAVIS, J. S. (1981). *Biochem. J.*, **197**, 309.
19. BERG, YU. N., LEBEDEVA, N. A., MARKINA, E. A. and IVANOV, I. I. (1965). *Biokhimiya*, **30**, 277.
20. O'SHEA, J. M., HORGAN, D. J. and MACFARLANE, J. J. (1976). *Aust. J. Biol. Sci.*, **29**, 197.
21. O'SHEA, J. M. and TUME, R. K. (1979). *Aust. J. Biol. Sci.*, **32**, 415.
22. SUZUKI, T. and MACFARLANE, J. J. (1984). *Meat Sci.*, **11**, 263.
23. IKKAI, T. and OOI, T. (1969). *Biochemistry*, **8**, 2615.
24. MACFARLANE, J. J. (1973). *J. Food Sci.*, **38**, 294.
25. KENNICK, W. H., ELGASIM, E. A., HOLMES, Z. A. and MEYER, P. F. (1980). *Meat Sci.*, **4**, 33.
26. MARSH, B. B. and LEET, N. G. (1966). *J. Food Sci.*, **31**, 450.
27. KENNICK, W. H. and ELGASIM, E. A. (1981). *Proc. 34th Annual Reciprocal Meat Conf.*, Am. Meat Science Assoc., Chicago, Illinois, p. 68.
28. RIFFERO, L. M. and HOLMES, Z. A. (1983). *J. Food Sci.*, **48**, 346.
29. SCHUMANN, M. S., KENNICK, W. H. and ROCK, D. F. (1982). *J. Food Sci.*, **47**, 864.
30. HORGAN, D. J. (1981). *Meat Sci.*, **5**, 297.
31. BRISKEY, E. J. (1964). *Adv. Food Res.*, **13**, 89.
32. MACFARLANE, J. J., MCKENZIE, I. J. and TURNER, R. H. (1982). *Meat Sci.*, **7**, 169.
33. HORGAN, D. J. (1979). *J. Food Sci.*, **44**, 492.
34. ELGASIM, E. A. and KENNICK, W. H. (1982). *Food Microstructure*, **1**, 75.
35. HORGAN, D. J. and KUYPERS, R. (1983). *Meat Sci.*, **8**, 65.
36. BOUTON, P. E., HARRIS, P. V., MACFARLANE, J. J. and O'SHEA, J. M. (1977). *Meat Sci.*, **1**, 307.
37. MACFARLANE, J. J. and MORTON, D. J. (1978). *Meat Sci.*, **2**, 281.
38. MACFARLANE, J. J., TURNER, R. H. and RATCLIFF, D. (1976). *J. Food Sci.*, **41**, 1447.
39. HERBERT, L. S. and SMITH, M. G. (1980). *CSIRO Food Res. Q.*, **40**, 65.
40. CUTHBERTSON, A. (1980). In: *Developments in meat science—1* (Lawrie, R. H. (Ed.)), Applied Science Publishers Ltd, London, p. 61.
41. ANON. (1980). *Nat. Provis.*, **182**, 18.
42. KOOHMARAIE, M., KENNICK, W. H., ELGASIM, E. A., DICKSON, R. L. and SANDINE, W. E. (1983). *J. Food Sci.*, **48**, 998.
43. ELGASIM, E. A., KENNICK, W. H., ANGLEMIER, A. F., ELKHALIFA, E. A. and KOOHMARAIE, M. (1982). *J. Food Sci.*, **47**, 861.
44. ELGASIM, E. A. and KENNICK, W. H. (1980). *J. Food Sci.*, **45**, 1122.
45. DAYTON, W. R., LEPLEY, R. A. and SCHOLLMEYER, J. V. (1981). *Proc. 34th Annual Reciprocal Meat Conf.*, Am. Meat Science Assoc., Chicago, Illinois, p. 17.
46. MORILD, E. (1981). *Adv. Protein Chem.*, **34**, 93.
47. BUSH, W. A., STROMER, M. H., GOLL, D. E. and SUZUKI, A. (1972). *J. Cell Biol.*, **52**, 367.

48. MACFARLANE, J. J., MCKENZIE, I. J., TURNER, R. H. and JONES, P. N. (1981). *Meat Sci.*, **5**, 307.
49. GREASER, M. L., WANG, S. and LEMANSKI, L. F. (1981). *Proc. 34th Annual Reciprocal Meat Conf.*, Am. Meat Science Assoc., p. 12.
50. VOYLE, C. A. (1969). *J. Food Technol.*, **4**, 275.
51. BOUTON, P. E., FORD, A. L., HARRIS, P. V., MACFARLANE, J. J. and O'SHEA, J. M. (1977). *J. Food Sci.*, **42**. 132.
52. DAVEY, C. L., KUTTEL, H. and GILBERT, K. V. (1967). *J. Food Technol.*, **2**, 53.
53. ROWE, R. W. D. (1974). *J. Food Technol.*, **9**, 501.
54. RATCLIFF, D., BOUTON, P. E., FORD, A. L., HARRIS, P. V., MACFARLANE, J. J. and O'SHEA, J. M. (1977). *J. Food Sci.*, **42**, 857.
55. BOUTON, P. E., HARRIS, P. V., MACFARLANE, J. J. and SNOWDEN, J. MCK. (1977). *J. Textural Studies*, **8**, 297.
56. BOUTON, P. E., HARRIS, P. V., MACFARLANE, J. J., O'SHEA, J. M. and SMITH, M. B. (1978). *J. Food Sci.*, **43**, 301.
57. SIMS, T. J. and BAILEY, A. J. (1981). In: *Developments in meat science—2* (Lawrie, R. (Ed.)), Applied Science Publishers Ltd, London, p. 29.
58. LOCKER, R. H. (1982). *Proc. 28th European Meeting of Meat Res. Workers*, Madrid, p. 117.
59. DAVEY, C. L. and GILBERT, K. V. (1976). *J. Sci. Fd Agric.*, **27**, 244.
60. KING, N. L. (1979). *Meat Sci.*, **3**, 75.
61. KING, N. L. (1984). *Meat Sci.*, **11**, 27.
62. BOUTON, P. E., HARRIS, P. V. and MACFARLANE, J. J. (1980). *J. Food Sci.*, **45**, 276.
63. MØLLER, A. J., BOUTON, P. E., HARRIS, P. V. and JONES, P. N. (1983). *J. Food Sci.*, **48**, 874.
64. BOUTON, P. E., HARRIS, P. V., MACFARLANE, J. J. and SHORTHOSE, W. R. (1982). *Meat Sci.*, **6**, 27.
65. ROBERTSON, J., BOUTON, P. E., HARRIS, P. V., MACFARLANE, J. J. and SHORTHOSE, W. R. (1984). *Meat Sci.*, **10**, 285.
66. SWIFT, C. E. and SULZBACHER, W. L. (1963). *Food Technol.*, **17**, 224.
67. HAMM, R. (1971). In: *Phosphates in food processing* (Deman, J. M. and Melnychyn, P. (Eds)), The AVI Publishing Co. Inc., Westport, Connecticut, p. 65.
68. OFFER, G. and TRINICK, J. (1983). *Meat Sci.*, **8**, 245.
69. MACFARLANE, J. J. (1974). *J. Food Sci.*, **39**, 542.
70. MACFARLANE, J. J. and MCKENZIE, I. J. (1976). *J. Food Sci.*, **41**, 1442.
71. MACFARLANE, J. J., MCKENZIE, I. J., TURNER, R. H. and JONES, P. N. (1984). *Meat Sci.*, **10**, 307.
72. HELLENDOORN, E. W. (1962). *Food Technol.*, **16**(9), 119.
73. PEARSON, A. M. and WOLZAK, A. M. (1982). *J. An. Sci.*, **54**, 1263.

Chapter 7

CONTAMINANTS IN MEAT AND MEAT PRODUCTS

A. RUITER

Faculty of Veterinary Medicine, University of Utrecht, The Netherlands

1. INTRODUCTION

Contaminants in food can be defined as undesirable compounds that are present in relatively small amounts, as a result of human action. This definition excludes additives which are intentionally introduced with the aim of improving some properties of the product under consideration (e.g. colour, flavour, texture, storage life). Compounds resulting from spoilage processes in meat or meat products are also excluded, and so are unwanted compounds that might be present naturally.

Figure 1 shows a scheme in which the most important groups of contaminants in meat and meat products are displayed. The following classification can be made:

(a) *Environmental contaminants.* These compounds are taken up by animals in their feed or drinking water and may be incompletely eliminated at the time of slaughter. In several cases these compounds tend to accumulate in some of the body tissues. Generally, the presence of environmental contaminants is not the result of animal treatment, nor are they introduced during processing.

A distinction can be made between compounds naturally present in meat but in much greater amounts as compared with non-contaminated products (e.g. heavy metals) and compounds not occurring in nature but synthesised or produced only by man (e.g.

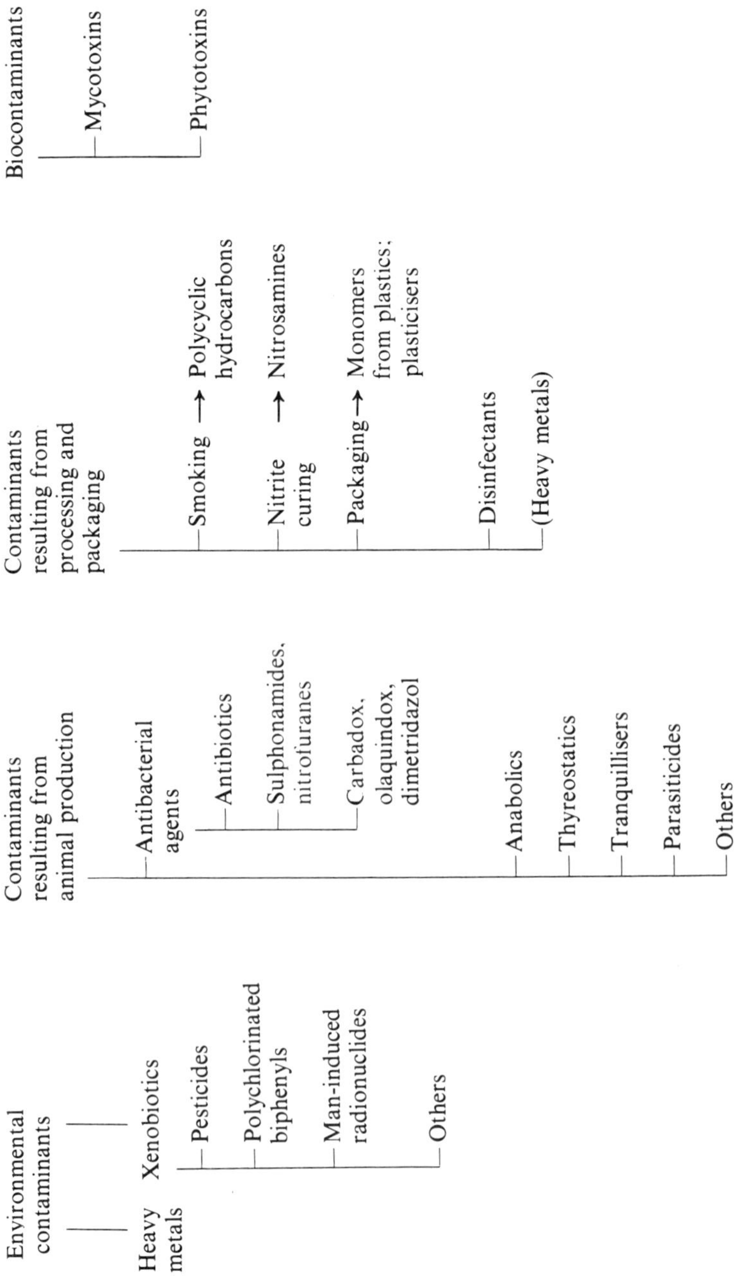

FIG. 1. Contaminants in meat and meat products.

chlorinated hydrocarbons, some radioactive isotopes). These latter are referred to as *xenobiotics* (*xenos* = strange, *bios* = life; strange to life).

(b) *Contaminants resulting from animal treatment*. During the growth and transport of slaughter animals, a number of drugs may be applied in order to cure the animals of infectious diseases, to promote feed conversion and growth, to minimise the effects of stress, etc. Under certain conditions, residues of these drugs or their metabolites may be present in meat or meat products.

(c) *Contaminants resulting from meat processing and packaging*. Some procedures may lead to undesirable compounds in the product (such as benzo[*a*]pyrene in heavily smoked products) or may give rise to the formation of unwanted compounds (such as nitrosamines in nitrite-cured products). Low molecular weight compounds present in wrapping materials may contaminate meat or meat products by simple migration.

(d) *Biocontaminants*. Some components of the feed of slaughter animals may be contaminated by toxins produced by certain fungi. Small amounts of these toxins, known as mycotoxins, are carried over from the feed to animal tissues, just as is the case with environmental contaminants. Since mycotoxins are not produced by human action, they do not fit the definition of food contaminants given above. However, the presence of these compounds in meat is also undesirable; they are therefore also dealt with in this chapter.

It must be stated that the term *residues* should preferably be reserved for remainders of compounds applied to slaughter animals in one way or another. Of course these applications themselves are intentional; the residues are not, but should be considered as unavoidable.

2. ENVIRONMENTAL CONTAMINANTS

Numerous substances are brought into the environment in some way or another by the actions of man. Some of these are notably resistant to chemical or biological degradation and will persist somewhere in the environment for a considerable period of time. In the long run, serious problems may arise from the presence of these compounds in the environment, in particular when they possess toxic properties. With regard to food of animal origin, special attention must be paid to those that tend to accumulate in the animal body. For that reason, the group of pesticides is considered here, followed by the chlorinated biphenyls and heavy metals. Finally, some words are devoted to artificial radioisotopes as far as they may be of importance to meat and meat products.

2.1. Pesticides

One of the properties of a suitable pesticide is a fair degree of persistence. This means that the compound should not be too susceptible to inactivation or degradation processes and that it should have a low volatility, so that it can exert its action for a considerable time. From this point of view, the well-known DDT (dichlorodiphenyl trichloroethane) is one of the most ideal insecticides. As a consequence, however, it is persistent in animals as well and, because of its high solubility in fat, will accumulate in adipose tissues. The major degradation product DDE (dichlorodiphenyl dichloroethene) is even more persistent and may be found in animal tissues for a notably longer time than DDT itself.

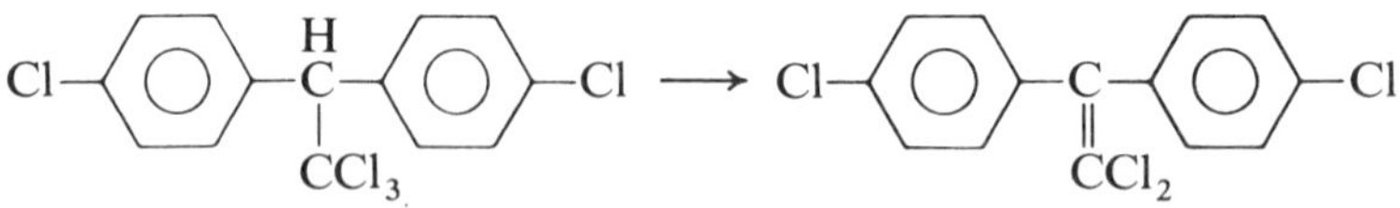

DDT DDE

Other organochlorine pesticides such as the cyclodiene group (aldrin, dieldrin, heptachlor, heptachlor epoxide, endrin, chlordane), the isomers of hexachlorocyclohexane (HCH) and the fungicide hexachlorobenzene (HCB) also tend to accumulate in body fat. (Hexachlorocyclohexane is often designated as benzene hexachloride (BHC), a term that might cause confusion with hexachlorobenzene, HCB.)

From the viewpoint of persistence the organochlorine pesticides can be arranged into four groups:[73] highly accumulating (HCB, β-HCH, and most pesticides of the cyclodiene group), moderately accumulating (endrin, heptachlor), low (α-HCH, γ-HCH) and very low accumulating (methoxychlor).

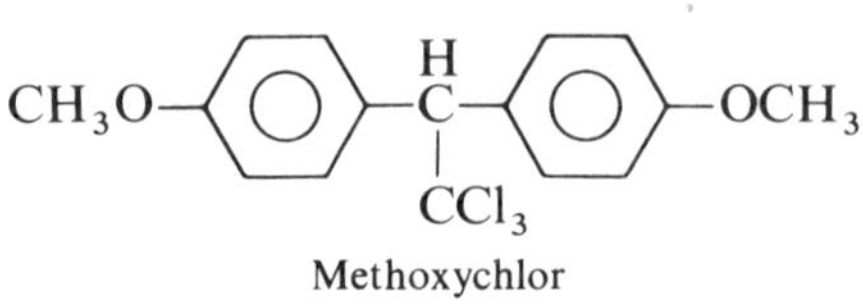

Methoxychlor

Accumulation ratios (tissue level to feed level) for abdominal fat of poultry are 10–30 for the first group,[73,125] 3–20 for DDT and its metabolites,[74] 3–7 for heptachlor and 1–3 for α- and γ-HCH.[73] In experiments with young beef steers, in which transfer of pesticides from feed to muscle tissue fat was studied,[156] figures are obtained from which comparable or slightly higher ratios can be calculated. Somewhat lower ratios were observed in adipose tissues of swine:[154,161] 8–11 for HCB, 1·6–3 for α-HCH, 2·1 for β-HCH, < 1 for γ-HCH, 2 for aldrin/dieldrin and 5–6 for the DDT group. Accumulation ratios may decline with increasing dietary levels, as is shown for HCB.[51]

For some organs, particularly the liver, the ratio may be much higher, values of more than 100 being reported.[36] (It should be noted, however, that it is somewhat problematic to define exactly the fat content of the liver. Apart from this, relating the amount of pesticides found to the content of alkane-extractable fat does not necessarily mean that the pesticides were present in this lipid fraction only.)

Excretion of chlorinated pesticides takes place via both faeces and urine.[10] Lactating cows excrete a considerable amount of these compounds by way of the milk; residues are also found in eggs of laying hens.

Information on half-life times is scarce. For laying hens, the following half-life times were observed in abdominal fat: 5–10 weeks for DDT and metabolites, the cyclodienes, HCB and β-HCH and 1·5–2 weeks for α- and γ-HCH.[73,74]

Although the organochlorine pesticides accumulate in the body lipids, this does not always mean that levels in animals are strictly related to the fat content.[101] Nevertheless, the fat content of meat determines to an important extent the amount of these compounds ingested by the consumer.

The acute oral toxicity of most organochlorine pesticides is not very high and decreases in the following order: aldrin/dieldrin > heptachlor/heptachlor epoxide > α-HCH > DDT > HCB. The toxicity of the latter compound is negligibly small; however, HCB is supposed to have carcinogenic properties and, because of its great persistence, contamination of food with HCB should be kept as low as possible.

Levels of organochlorine pesticides in meat are well below the maximum residue levels set up in various countries. From current analyses performed in the Netherlands[139] it can be concluded that the level of the DDT group averages 50 μg/kg of extractable fat, which means that not more than a few μg of DDT are present in 1 kg of meat. This is in accord with figures given for beef;[101,102] however, distinctly higher levels are reported for pork and much higher levels for horse, chicken, turkey and rabbit meat (up to 60 μg/kg of wet weight in chicken meat[101]). In current Dutch analyses,[139] higher levels in poultry were also occasionally found; this also holds for other pesticides.

The cyclodiene group as a whole shows levels which are usually below 20 μg/kg on a fat basis; the same holds for HCB.[139] In another publication, however, mean HCB levels of 50 μg/kg have been reported.[40]

The HCH isomers in meat show values in the range of 10–100 μg/kg on a fat basis.[139] γ-HCH, apart from its occurrence in the environment, is still used in animal husbandry and for this reason residue levels are often in excess of those of the other isomers. Occasionally, β-HCH may occur at higher levels because of its great persistence, which far exceeds that of the other isomers. Pesticide levels can be reduced, to a certain extent, by cooking.[94,109,127,128] The losses were primarily due to leaching, although some compounds (γ-HCH, heptachlor epoxide) underwent some destruction by heating. The HCH isomers and DDT may be subject to partial degradation by yeasts and micrococci,[34,104,115] which results in reduced levels in fermented sausage as compared with the starting material.

Owing to regulations in several countries the levels of most organochlorine pesticides showed decreases in the period 1970–80. Considering the years 1974–79, decreases in beef were observed for the DDT group, the cyclodiene group and for HCB, but not for the HCH isomers.[8] Pesticides other than organochlorine compounds (e.g. organophosphates, methylcarbamates, pyrethroids) are considered to lead to no appreciable residue levels because of their much shorter persistence. Methyl carbamates, for example, are degraded for the greater part within 12 h after oral administration, while an equivalent of 80 % of the original dose is excreted via the urine and from 0·5 to 15 % via the faeces within 24 h.[27] As for the pyrethroids, these are, as a rule, rapidly metabolised in mammals;[14] occurrence of appreciable amounts in tissues of slaughter animals is improbable, as far as can be ascertained at present.

2.2. Polychlorinated Biphenyls

Polychlorinated biphenyls (PCBs) are synthetic compounds with excellent chemical and thermal stability. They are used for many purposes, e.g. isolation in transformer installations and capacitors, lubrication, impregnation of paper and as plasticisers in synthetic polymers. The general formula can be written as follows:

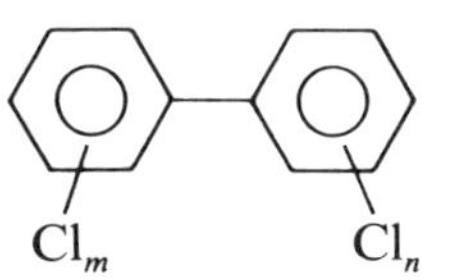

$(m, n = 1, 2, 3, 4 \text{ or } 5)$

Commercial PCB preparations are complex mixtures of these components, as more than 50 different PCBs may be present. PCBs are extremely persistent and, because of their high solubility in fat, tend to accumulate in adipose tissues of man and animals just as do organochlorine pesticides. As it has been shown that these components, despite their chemical inertness, are far from harmless (toxic properties have been amply described; teratogenic and tumour-promoting effects were demonstrated and induction of tumours by PCBs seems probable[12]), the presence of these compounds in the environment presents a real problem.

For several reasons, PCBs accumulate particularly in fish. Fish from certain inland waters in Western Europe may show PCB levels exceeding 10 mg/kg on a fat basis. The concentrations are considerably lower in meat and meat products. Sümmerman *et al.*,[142] reviewing the situation in the Federal Republic of Germany, have given some indicative levels (in μg/kg) for total PCBs in some foodstuffs:

Low-fat food components of vegetable origin such as cereals or potatoes	5
Eggs	30
Vegetable fats	50
Fats from land animals	150
Milk fat	300
Fish fat	10 000

In another publication[8] levels of 200–300 μg/kg in beef and 100 μg/kg in pork (on a fat basis) are mentioned. These levels are expressed as Clophen A-60, a commercial PCB preparation containing 60 % Cl. A Dutch survey reports average values of 150 μg/kg for beef fat and 70 μg/kg for pork fat.[139]

Fat from land animals, including the fat present in meat, contributes about one-quarter to the total amount of PCBs in food.[142] There are no indications that the amount of PCBs daily ingested (15–35 μg) will cause direct harm to the consumer, but because of the properties of these compounds, a further increase in the levels present in food is undesirable. It should be mentioned that in several countries measures are being taken, or will be taken, to lower the production of PCBs.

Accumulation ratios vary with individual PCBs. In a recent publication on PCB accumulation in chicken tissues many data concerning these individual components have been presented.[52] One of the conclusions that can be drawn is that less chlorinated biphenyls (4 Cl atoms per molecule) tend to lower accumulation ratios than more chlorinated biphenyls (6 and 7 Cl atoms per molecule). It was found earlier that the transfer, from feed to milk, of lactating cows is strongly dependent on the degree of chlorination.[151] However, PCBs low in chlorine will produce more reactive metabolites in a given interval of time[12] and are more toxic as compared to more chlorinated biphenyls. Replacement of the latter compounds by less chlorinated biphenyls will therefore not solve the problems generated by the use of these compounds.

2.3. Heavy Metals

Heavy metals are natural constituents of meat that have always been present in all tissues of animal origin. As a result of utilisation by man, however, larger amounts have spread over the Earth's surface, which has also caused a gradual increase in some of these elements in animal tissues. This holds particularly for non-essential elements such as lead, cadmium and mercury which, as far as is known at present, do not fulfil any specific requirement in the animal body. Because of their toxic properties and their tendency to accumulate in man and animals, these three elements deserve special attention. Many investigators have determined concentrations of these elements in edible animal tissues such as muscle, liver and kidney. Generally, the concentrations in liver and kidneys are considerably higher than in muscle tissue. In other parts occasionally eaten by man (e.g. spleen, heart, brains, lungs) these concentrations will be lower, as is demonstrated for lead and cadmium.[54] The risk caused by other toxic elements that may be present in measurable amounts in meat and meat products is considered to be small; therefore, only minor attention is paid to these elements here.

2.3.1. Lead

Of the lead in food, 8 % or less is taken up from the human gastrointestinal tract;[11] in young children, however, this uptake is considerably higher. Of retained lead, more than 95 % is deposited in bone tissue; deposition and mobilisation are regulated in the same way as for calcium.[164] Other target organs are the liver, kidney, brain and bone marrow. It has been proved that lead interferes with haemoglobin synthesis and is able to inhibit several enzymes.

The lead intake for an adult weighing 60 kg should not exceed 3 mg/week.[164,165] In the Federal Republic of Germany guideline values for lead in food have been established:[72] 0·3 mg/kg for the meat of cattle and swine and 0·8 mg/kg for livers. In the meat of these animals the contents are usually below 0·05 mg/kg,[40,54,80,82] whereas sheep have slightly higher values.[54,77,82] In the meat of horses the lead content occasionally exceeds the guideline, which is in keeping with the phenomenon that aged horses tend to accumulate lead in their muscle tissues.[54]

Extremely high lead concentrations are sometimes found in game, which are caused by absorption of shot left in the body.[55] This is particularly the case when the animals were hit by fragmentation shot.[108]

The 0·8 mg/kg level is reported to be frequently exceeded in the livers of slaughter animals;[80] in another study, however, no values exceeding this level were observed.[54] This also holds for kidneys.[149] In an Australian study the contents measured in liver vary from non-detectable to 0·85 mg/kg, with a mean value of 0·14 mg/kg.[39]

It is clear from various publications that the lead content of animal tissues depends on the local contamination with this element.[3,63,64,77,80] Contact with metal surfaces may contribute to this content, as also may the addition of lead-containing additives such as certain spices[55] and salt.[53]

2.3.2. Cadmium

Cadmium accumulates strongly in the kidney and, to a lesser extent, in the liver as well. It has an extremely long biological half-life in man in the order of 20–30 years.[164]

From the cadmium present in food, 5–10% is estimated to be taken up. Being absorbed as Cd^{2+}, it is bound in the liver to metallothionein (a polypeptide in which about one-third of the component amino acids consists of cysteine) and stored in the kidney as Cd–thionein. Kidneys containing more than 200 mg Cd per kg may exhibit clinical anomalies such as morphological abnormalities in the tissue structure (i.e. tubular damage) and increased excretion of proteins, amino acids, glucose and calcium.[37,42] Other symptoms of chronic cadmium poisoning are emphysema of the lungs, anaemia and insufficient bone mineralisation (osteomalacia) as a result of increased calcium excretion.

In the kidney and the liver, and possibly other tissues as well, at least a part of the cadmium present is bound to metallothionein. It should not be forgotten that cadmium ingested in this form is transported directly to the kidneys[15] and that Cd–thionein must be considered more toxic than Cd^{2+}.[159] Insufficient data are available, however, concerning the uptake of Cd–metallothionein complexes by the gastrointestinal tract. The cadmium content of muscle tissue is nearly always below 0·05 mg/kg and is mostly below 0·01 mg/kg.[40,54,77,81,82] Cadmium concentrations may be high in kidney and liver tissue.[39,78,82] These concentrations are related to age and to concentrations in feed.[54,81,82,168] A relation between the Cd concentration in the kidney and age was also observed in game.[62]

Cadmium concentrations in horse kidney are sometimes extremely high.[54,61,133] This is explained by the fact that horses are not very young when slaughtered. Kidneys of aged animals (horses, game) are therefore unsuitable for consumption.[61] In most slaughter animals, however, Cd levels in the kidney and liver do not exceed 0·8 and 0·5 mg/kg, respectively. Finally, it should be noted that, in normal diets, meat, fish and poultry contribute only 5% to the total amount of cadmium ingested.[41]

2.3.3. Mercury

Mercury levels in meat and meat products do not give rise to much concern. The contents in muscle are in the order of 0·005 mg/kg.[40,77,82,102] Those in livers may exceed 0·01 mg/kg but contents higher than 0·05 mg/kg are exceptional.[39,78,82] Kidneys show somewhat higher levels which in some cases exceed 0·10 mg/kg.[39]

2.3.4. Other Elements

In general, it can be said that carry-over of heavy elements to muscle tissue is low. Copper levels, for example, remain constant in muscle tissue even when copper is fed to pigs in amounts of 500 mg/kg of feed.[76] The livers, however, accumulate copper, and so do the kidneys, although to a lesser extent. The rate of excretion from these organs varies from element to element: cadmium is hardly excreted at all, while arsenic is rapidly excreted.[22] The presence of traces of arsenic, although a rather toxic element, should not be considered to be harmful or hazardous in meat.[60] Contents in meat are generally below 0·01 mg/kg.[40]

As to the other elements, thallium can be mentioned as a heavy metal to which some attention has been paid recently as a possible element of concern.[79] More data should be made available, however, before any conclusion is drawn.

2.4. Radionuclides

In animal tissues a natural radioactivity of 100–200 Bq/kg* always exists. This activity is due mainly to the presence of the natural isotopes ^{40}K (about 40% of the total activity), ^{14}C (about 25%) and ^{87}Rb. The contributions of other isotopes (^{3}H, ^{226}Ra, ^{210}Po) are very small.

About 200 radioactive isotopes have originated from man-induced nuclear reactions. Most of them play no role in meat and meat products. For them to be of concern, the following requirements must be fulfilled:

* 1 Becquerel (Bq) = 1 disintegration/s = 27 pCi.

generation in appreciable amounts during nuclear reactions; a sufficiently long physical half-life time; rapid fall-out from the atmosphere; good absorption by plants; good absorption from plants or vegetable feed by animals; not too high an excretion ratio. In fact only the isotopes ^{137}Cs, ^{90}Sr and ^{3}H meet these requirements. As ^{90}Sr is found for the greater part in the bones (and, in the case of lactating animals, in the milk), only ^{137}Cs and ^{3}H are of interest to meat and meat products. ^{137}Cs in particular is measured in meat for monitoring purposes.

As a result of experiments with nuclear weapons in the atmosphere, some contamination of meat with ^{137}Cs has been observed. During the years 1962–64 values in the range of 15–30 Bq/kg were found.[6,16] Over the years this contamination gradually decreased and it now amounts to no more than 0·7 Bq/kg on average. This radiation still originates from experiments with nuclear weapons performed in the past; the overall contribution from nuclear reactors is negligibly small.

In the Netherlands the average annual dose of radiation due to man-induced radioactivity was estimated to be 10 μSv* in 1981.[16] In this value the exposure to radiation from ingested food was included. Consumption of meat and meat products may contribute 0·1 μSv. As the annual dose due to the natural background is in the order of 2000 μSv[153] it is obvious that meat and meat products do not make a significant contribution to the total exposure to ionising radiation.

3. CONTAMINANTS RESULTING FROM ANIMAL TREATMENT

Obviously it is not feasible to discuss here all preparations that may be applied to slaughter animals during their life-span for various reasons, as many thousands of substances are included in this group. In principle, all preparations administered to animals will lead to residues in the tissues when the time between administration and slaughter is too short. However, many of these compounds given orally to the animal are rapidly metabolised and excreted, so that they reach the muscle tissue in very small quantities only; even in the liver and the kidney, the concentrations decrease rapidly after stopping the administration of the compound in question.

* 1 Sievert (Sv) = 100 rem.

Preparations administered parenterally may remain in the site of injection for a considerable time, especially when the application is by way of intramuscular, subcutaneous or intracutaneous injection or via an implant. Care should be given to this site with respect to the presence of residues derived from the preparations used. In some cases it is recommended to use sites that are not consumed.

For compounds used in animal treatment, withdrawal times should be established in order to ensure that residues are below an acceptable or a measurable concentration level. This holds particularly for some antibacterial agents and also for certain synthetic anabolics. In general, compounds of a lipophilic character are excreted much more slowly than polar compounds. In some cases excretion is enhanced by conversion into more polar compounds or by conjugation with a polar compound such as glucuronic acid. These conjugates, however, may be reabsorbed from the intestinal tract after hydrolysis by the intestinal flora (the entero-hepatic cycle), thus extending the time necessary for complete excretion.

It should be noted also that proper observance of withdrawal times is not always a guarantee of the absence of residues in meat. Withdrawal times are usually established on the basis of experiments with healthy animals; relative to the health status of the animal(s) under consideration, the withdrawal time has to be multiplied by a certain factor. In most cases, however, this factor is not well known and arbitrary factors have to be used in order to ensure sufficient removal of residues. Withdrawal periods are also dependent on the pharmaceutical formulation in which the compound is present, and on the mode of administration.

Another problem is that it is difficult to guarantee that certain additives are no longer present in the feed some time before slaughtering has been planned. Both in feed factories and in farmhouses, it is feasible that feed free of additives such as antibacterial drugs could be contaminated with feed containing these compounds and that traces will be present in the animal at the time of slaughter. For that reason, it is of great importance to impose strict control in both factories and farmhouses to minimise such contamination.

3.1. Antibacterial Agents

Antibacterial drugs, including the antibiotics, may be applied to slaughter animals for curative, preventive and nutritive purposes. In many cases these are administered as a feed additive, by way of medicated feeds or via the drinking water. Parenteral administration is practised mainly in cases of acute disease. If proper withdrawal times cannot be observed (e.g. in an emergency slaughtering), traces of antibacterial agents may be present in the tissues. Some antibacterial drugs are discussed briefly below.

The *sulphonamides* present an important group of antimicrobial drugs which can be formulated as follows:

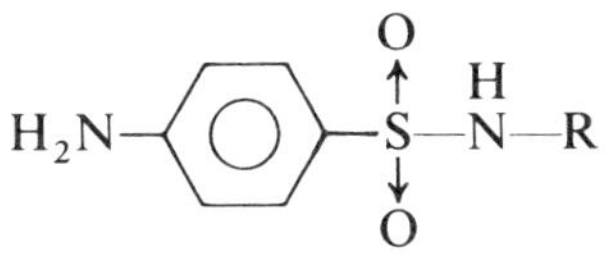

where R stands for a specific group, mostly of a heterocyclic nature. Toxic effects of sulphonamides have been reported,[134,160] but in general the toxicity of these compounds is comparatively low. In the animal body the sulphonamides are partly metabolised and/or conjugated. Acetylation plays an important part in the overall metabolism; acetylated sulphonamides have lost their antibacterial action. Withdrawal times of two or more weeks are necessary[150] to ensure that sulphonamide residues are below a level of 0·1–1 mg/kg in animal tissues.

Nitrofuranes are particularly used to prevent poultry, swine and calves from getting *Salmonella* and *E. coli* infections. The most widely used representative of this class is furazolidone:

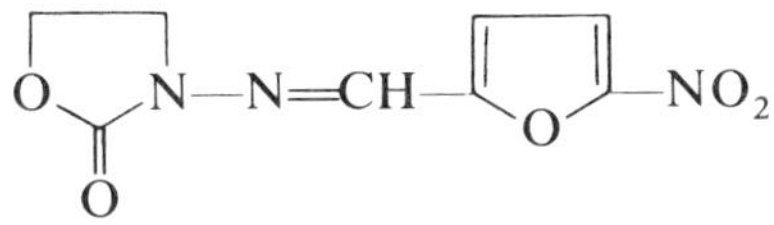

Furazolidone is rapidly metabolised in the animal body, but the metabolic pathways, and the nature of the products formed, are almost completely unknown. It has been found that many 5-nitrofurane derivatives show carcinogenic properties. This also applies, in all probability, to furazolidone and to metabolites in which the 5-nitrofurane structure is still intact. Mutagenicity has been proved.[99,143] Experiments with mice and rats show an increased incidence of different types of carcinomas following administration of furazolidone. However, these data were available only from secondary sources and, according to the principle adopted in the International Agency for Research on Cancer (IARC) Monographs Programme, no further evaluation of the carcinogenicity of furazolidone for experimental animals could be made from these.[65]

Few data on residue levels in slaughter animals are available. For poultry and swine a withdrawal period of 3 weeks has been recommended.[5]

The *tetracyclines* belong to the antibiotics most frequently applied in veterinary practice at present. These components show a very broad antimicrobial spectrum, which includes Gram-positive and Gram-negative bacteria, *Spirochaeta*, *Rickettsia* and *Mycoplasma* species. The molecules consist of four condensed six-membered rings which are partly unsaturated and to which several hydroxyl, carbonyl and methyl groups are attached. Tetracyclines are able to form stable chelates with bivalent cations such as Ca^{2+} and Mg^{2+} and with many other substances. Because of the strong affinity to Ca^{2+}, these compounds tend to be incorporated into the bone tissue, from which they are released slowly.

Various toxic effects of large doses have been described, but mutagenic or carcinogenic properties have never been observed. The metabolism is largely unknown. In spite of the slow release from bone tissue mentioned above, excretion may be complete within 3–5 days.[1,2,33,75] In human therapy the strong affinity towards Ca^{2+} may lead to affection of the teeth, but such an effect is unlikely from the residual amounts that may be present in animal tissues.

Chloramphenicol, like the tetracyclines and the sulphonamides, is active against both Gram-positive and Gram-negative bacteria. The structure is most unusual for a biological product in that it includes a nitrophenyl group and a dichloroacetic acid moiety. The metabolism is somewhat better known than that of most other antibiotics.[46] The main product is a conjugate of the parent compound with glucuronic acid. In cattle most of the chloramphenicol administered is excreted in this form with the urine.[90]

For cattle and other husbandry species the toxicity of chloramphenicol is low so that it is used freely in veterinary medicine. As chloramphenicol is rapidly excreted, residues in edible tissues are usually low. Apart from this, it should be pointed out that routine microbiological methods for screening residues of antimicrobial agents in food of animal origin are not very sensitive towards chloramphenicol.

Application of chloramphenicol to man may cause adverse effects such as aplastic anaemia or 'grey baby syndrome' in infants, both with low frequency but with high mortality. There are indications that these effects are relatively dose-independent. For this reason, residue levels should be as low as possible.

Antibiotics of the *penicillin* group, as a rule, are rapidly excreted. As to metabolism, the first degradation step is a hydrolytic cleavage of the β-lactam ring resulting in the inactive penicilloic acid:

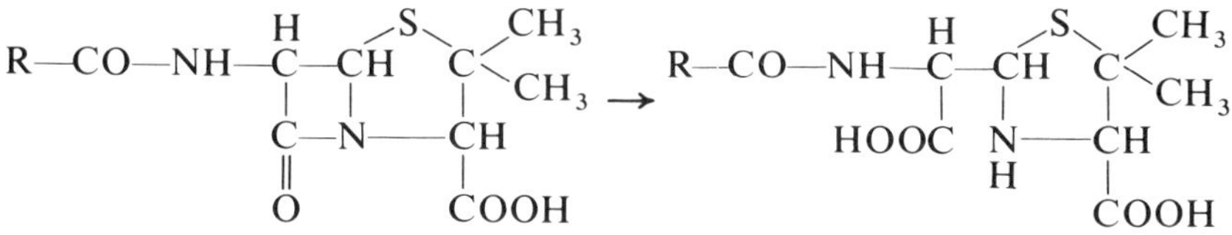

Penicillin Penicilloic acid

Penicillins can be sensitively detected by microbial growth inhibition tests; this has contributed strongly to the introduction of low tolerance levels. As such, penicillins are among the safest antimicrobial agents, as their toxicity is low and mutagenic or carcinogenic properties have never been demonstrated. It has been frequently stated that residues of penicillin in food may elicit allergic reactions in hypersensitive consumers. In a prospective investigation, however, no clinical symptoms were observed.[93]

Carbadox® is a synthetic feed additive applied to swine in order to prevent shigellosis and bacterial enteritis. As the addition of Carbadox to feed results in faster growth and improved feed conversion, this compound is often indicated as a growth promoter. Carbadox is rapidly metabolised and excreted, the main metabolite being quinoxaline-2-carboxylic acid:

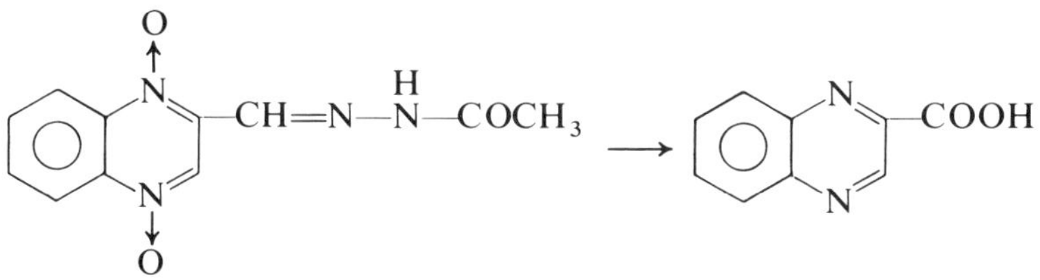

Carbadox shows mutagenic activity[38] and is generally considered to be carcinogenic as well. Residues should therefore be absent in animal tissues. To achieve this, Carbadox should be withdrawn from the feed of piglets at the age of 15 weeks at most, and at least 4 weeks before slaughter.

3.2. Anabolics

Anabolic compounds, either natural or xenobiotic, influence the metabolic functions of slaughter animals such as cattle and sheep, resulting in an increased growth rate and also an increased deposition of proteins in the muscle mass. Anabolics have been applied in animal production for more than 30 years in order to improve food conversion,[157] especially in ruminants.

Nearly all the anabolics used in practice also act as sex hormones. Many of them possess oestrogenic properties, while others are androgens or gestagens. It is not unusual for combined preparations (e.g. an oestrogen plus an androgen) to be applied. Some of the compounds used are naturally occurring hormones and belong to the class of steroids, a typical example being 17β-estradiol. Others are modified natural hormones such as esters which decompose in the body into the natural parent compounds, modifications in which methyl, methylene, ethynyl, acetoxy or other groups are introduced and compounds in which extra double bonds have been created. A third class comprises synthetic anabolics such as diethylstilboestrol (DES) and zearalanol (Zeranol®), a component obtained by chemical reduction of the mycotoxin zearalenone (see Section 5.1). Naturally occurring anabolics (sometimes improperly designated as endogenic anabolics) are usually inactive when ingested orally and are administered parenterally or by means of implantation. Modifed natural and synthetic anabolics often resist breakdown in the gastrointestinal tract and may be administered orally; however, anabolics belonging to these classes are applied by injection or implantation as well. Residues of anabolics may be present in edible tissues, but apart from the site of injection or implantation, the levels are generally low and seldom exceed the low μg/kg range. The anabolics may be partially present as metabolites or as conjugates.

Implantation of a naturally occurring oestrogen such as 17β-estradiol in steers results in somewhat higher concentrations of estrogen in edible tissues, naturally occurring values of 20–40 ng/kg being roughly doubled; however, these concentrations remain far below those attained in pregnant cows.[56]

The use of controlled implants of 17β-estradiol only theoretically increases human exposure to oestrogens. Amounts ingested with cow's milk are usually higher than those from the tissues of implanted animals.[163] In their turn, concentrations in milk are negligibly small when compared to the amount of oestrogen endogenously produced every day by the consumer, even the young one.

As for the residues of xenobiotic anabolics, it was found in several investigations that concentrations in edible tissue are also low. For Zeranol, Janski[68] found a level in implanted cattle of below 0·2 μg/kg, which means that it could not be distinguished from that of non-implanted controls. Levels in the kidney, adipose tissue and the liver were in the order of 0·5–1 μg/kg. These levels are well below the value of 8·3 μg/kg provisionally calculated for the tolerable level of Zeranol in meat.[97]

Comparable values were found for the androgenic compound Trenbolone® in implanted heifers, the highest values being found in the liver. All values were below the maximum of 1·7 μg/kg proposed by Van Logten *et al.*[97] However, experiments with ^{3}H-labelled Trenbolone suggested much higher residue values (up to 25 μg/kg). It was found that the major part of the radioactive material could not be extracted from the matrix, which may indicate that either Trenbolone or one or more metabolites were bound to proteins or protein-like material,[132,136] which strongly reduces bioavailability.

Melengesterol®, a modified progesterone applied orally as its acetate to heifers, occurs in tissues of treated animals in concentrations usually below 10 μg/kg.[87,110] A value of 25 μg/kg is mentioned as a tolerance limit.[110]

The use of diethylstilboestrol (DES) as an anabolic agent is prohibited in most countries because of its carcinogenic properties. Sensitive methods have been developed for the estimation[7] and identification[24] of DES in the urine of cattle, in which the concentrations are considerably higher than in muscle tissue. At present, DES can be detected at a level of 1 μg/kg of urine, which roughly corresponds to a concentration in muscle tissue of 0·02 μg/kg.[67]

The major risk of administration of anabolics to slaughter cattle is the improper use of these compounds (application of high doses; careless injection or implantation, leading to possible consumption of the injection or implantation site; use of DES or related compounds). On the other hand, there is little cause for concern when a natural oestrogen such as 17β-oestradiol is used properly in animal production, and similar considerations apply to progesterone and testosterone.[69] As for the proper use of natural hormones, the presence of minute amounts of these compounds or their derivatives in edible tissue does not seem to be of any significance. For xenobiotic anabolics such as Zeranol and Trenbolone, application will remain a subject for discussion for the time being.

3.2.1. Thyreostatics

Substances inhibiting the thyroid function, also known as thyreostatics, allow a considerable increase in the live weight gain of slaughter animals. This is not due to an increased growth rate or to a deposition of muscle proteins, but mainly to an augmented water retention in the muscle mass and an increased filling of the gastrointestinal tract.[162]

Thyreostatics, if applied, are fed daily to adult cattle; residues of the compounds most frequently used (methyl thiouracil (MTU), tapazole) may be found in the liver, kidney and muscle tissue, and naturally at the highest level in the thyroid gland.[9] These residues are probably not of great harm to the consumer but, because of their interference with a hormonal action, the use of thyreostatics is prohibited in many countries. Another reason for the prohibition is that the increase in the weight of slaughter animals is due only to water retention and not to improved feeding efficiency. The use of thyreostatics can therefore be regarded as improper practice.

Supervision is focused upon the use of thyreostatics as such and not primarily upon the determination of residue levels in tissue, although reliable methods for that purpose are available.[9,88] Application of thyreostatics to slaughter animals can easily be detected by a decreased level of thyroid hormones in blood plasma[116] or by a weight increase of the thyroid gland.[155]

3.3. Tranquillisers

In meat-producing animals such as pigs, sensitivity to stress is often a serious problem. Insufficient capacity to adapt to excessive situations, e.g. high ambient temperatures and transport to the slaughterhouse, is in part due to genetic factors and in part is the result of intensive breeding. Stress is manifested by increased motor activity and emotional reactions, accompanied by symptoms such as disturbances in respiration, blood circulation, hyperthermia, cyanosis and muscle rigidity. As a result of metabolic changes in the muscle, stress often results in a decrease in meat quality (pale, soft, exudative (PSE) meat or dark, firm, dry (DFD) meat).

The phenomenon of stress has recently been reviewed by Lister *et al.*;[95] that review also discusses the biochemical changes that occur during development of stress. Losses in animal production by stress-induced death may amount to up to 0·37% in pigs.[18] Emergency slaughtering is necessary to a somewhat greater extent.[25]

The incidence of stress during transport of slaughter animals can be decreased by improvement of transport conditions. Moreover, the selection of a pig race less sensitive to stress has proved successful in combating the problem.[32] Parenteral administration of tranquillisers is another method of diminishing the losses caused by stress during transport. Azaperone (Stresnil®), a butyrophenone derivative, is frequently used for this purpose. Being a neuroleptic, it reduces emotional reactions, motor activity and hyperthermia. Glycolysis and losses of ATP are also reduced.

As tranquillisers are generally administered shortly before slaughtering, i.e. just before transportation, a real chance exists that residues of these compounds may be present in the animal after it has been killed. With regard to azaperone, injection of the recommended dose (0·4 mg/kg of body weight) may lead to concentrations in muscle tissue of 0·05 mg/kg after 15 min to 1 h,[123] which is far below the proposed maximum level of tolerance.[124] In the kidney, this level (1 mg/kg) may be reached at 15 min after administration but falls to 0·3–0·4 mg/kg within 1 h.[123] Concentrations in the liver are lower than in the kidney.

The point of injection may give rise to some concern, as amounts of up to some mg have been observed there 1 h after injection.[123] Some evidence has recently been obtained that azaperone has weakly mutagenic properties.[121] However, the acute toxicity seems to be low, a no-effect level of 1·25 mg/kg/day being observed in experiments with dogs and rats.[124]

Azaperone is rapidly reduced to azaperol in the body; this metabolite still has some neuroleptic activity:[122,123]

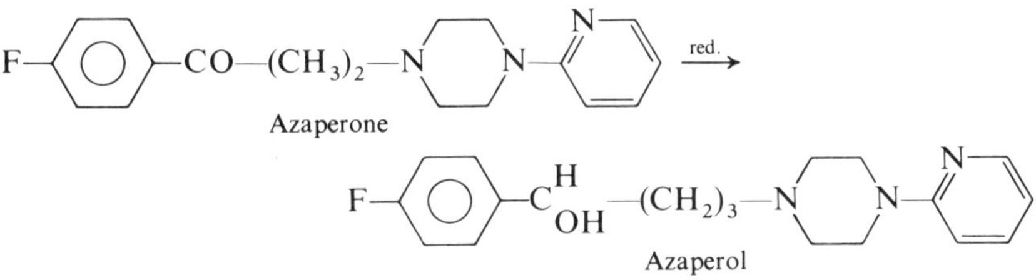

Oxidative breakdown, which is also observed,[58,59] leads to inactive products. Data on residues of other tranquillisers are scarce. Residue levels of the neuroleptic propiopromazine in slaughtered animals are comparable to those of azaperone.[113]

The presence of tranquillisers in slaughter animals does not seem to give rise to great public health problems; however, some care should be taken with regard to the sites of injection.

3.4. Parasiticides

In modern animal husbandry it is necessary to use compounds that are able to combat parasites in or on the animal. Many different preparations containing a variety of compounds are available; it has been stated that no residue problems are to be expected, as these preparations will not be applied shortly before slaughter,[6] and because most compounds are metabolised and/or rapidly excreted. However, attention should be paid to the fact that some *ecto*-parasiticides may penetrate the skin and, in cases where these compounds are lipophilic, they are retained for some time in adipose tissues.

This holds especially for lindane (γ-HCH), which may be found in these tissues several weeks after dermal application. Even some organophosphates which are excreted rapidly after oral administration can be found in the body fat of steers at 10 days after administration.[166] It should be taken into account that uptake via the skin may be considerable and that amounts entering the body by this route may exceed amounts ingested orally.

4. CONTAMINANTS RESULTING FROM MEAT PROCESSING AND PACKAGING

4.1. Polycyclic Aromatic Hydrocarbons

A classic example of contamination of meat during processing is that with polycyclic aromatic hydrocarbons (PAHs) as a result of some types of smoking processes.

PAHs are formed during pyrolysis of organic matter. Although nearly all the PAHs now found in foodstuffs originated from human activities, these compounds were present on Earth long before man appeared on the scene, as they are generated during fires caused by thunderstorms or volcanic activities. PAHs are present for the greater part in tarry droplets and soot particles generated in incomplete burning processes. They are characterised chemically as compounds containing two or more condensed aromatic ring systems, usually benzene ring structures.

Some PAHs are potent carcinogens and are the cause of the carcinogenic properties of soot and tar which have been recognised for a long time. In particular, the component benzo[*a*]pyrene (BaP) is known as a carcinogen:

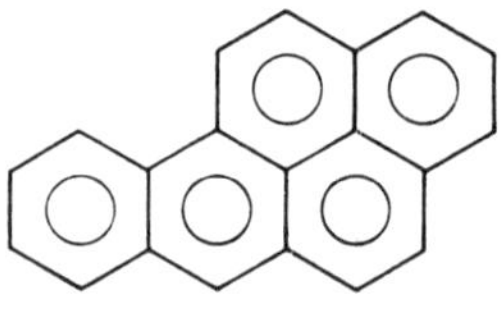

Benzo[*a*]pyrene

However, many more PAHs present in soot and tar are able to cause tumours. It has been suggested that BaP represents only 10% of the carcinogenic potency found in this material.[119] Others mention a range of 1–20% of total carcinogenic PAHs.[4] The effect of these compounds is additive. There are no indications for synergism, but antagonism is certainly not excluded. Antagonism has been reported in the past, but recent information is lacking.[119]

Many foodstuffs, particularly those of vegetable origin, contain measurable amounts of PAHs as a result of plant exposure to the atmosphere which may be polluted with particulate matter containing these substances. Raw meat does not contain PAHs,[96,117] but they will be present after smoking or grilling. The PAH content of smoked or grilled meat depends on the way in which these foods are prepared. Smoke can be described as an aerosol in which the dispersed phase consists of tarry droplets that contain nearly all the PAHs present. In a conventional smokehouse where the meat is hung directly over the smouldering fire, notable amounts of this disperse phase are deposited on the surface of the product. In modern smoke installations the smoke is not generated in the smoke kiln itself and has to be led into it by means of a duct. On its way to the product, the smoke loses a great part of its dispersed phase, especially the larger particles. These losses are mainly caused by the impact of these sticky particles against the walls of the duct and in the ventilator (if present) and by the fact that the smoke is cooled down. Cooling of smoke results in an increase in particle size which makes removal easier. Meat smoked in this way contains far fewer PAHs than, for example, home-smoked meat.[118,144]

By the use of experimental smoke generators, it has been established that the amount of PAHs produced is a function of the temperature in the zone where the smoke is generated.[148] At 400 °C there is little production of PAHs from wood, but at higher temperatures PAH production increases markedly. In a smouldering fire, however, the temperature varies from place to place and will be much higher than 400 °C in most regions; therefore, PAH formation cannot be avoided.[145] Newer smoke generators, e.g. those in which wood is pyrolysed by means of superheated steam, produce smoke at low and controlled temperatures. Moreover, at moderate combustion temperatures the production of aromatic components such as substituted phenols is higher.[120]

Sausage casings are an effective barrier against PAH penetration into the product. It has been established, however, that there is little penetration of these compounds at all and that the greatest amount is found in the outer layer.[126,146] Treatment of meat and meat products with smoke preparations ('liquid smoke') allows control of PAHs in the final product, as smoke preparations low in PAHs can easily be obtained.

As for the smoking of meat and meat products, the technological procedure should be performed in such a way that deposition of PAHs is as low as possible. Based upon a recommendation of Tóth and Blaas,[147] the Federal Republic of Germany in 1973 established a maximum level of 1 μg of BaP per kg of smoked meat or meat product. This tolerance level was not based upon toxicological considerations but on the statement that, by employing modern smoking techniques, products with a BaP content well below this level can easily be obtained. In home-smoked meat and in traditional products such as 'black smoked' ham, this content is often above 1 μg/kg and may reach levels exceeding 10 μg/kg.[118,144]

Another source of PAHs in meat is grilling. In charcoal-grilled meat, BaP contents are found which far exceed the 1 μg/kg level and in some cases may reach levels of more than 100 μg/kg.[26,89]

A proportional relation was found between the amount of BaP and the fat content of the meat.[26] Lyinsky and Ross,[100] who observed this phenomenon earlier, explained it by the dripping of fat on to the hot coals where it is subjected to pyrolysis; BaP and other carcinogens formed are in part deposited on the meat surface. If little fat is available to drip from the meat, this deposition is obviously much smaller.

Apart from this, other carcinogens may be formed on the meat surface, in particular when protein-containing material comes into direct contact with the fire. It is known that some compounds originating from pyrolysis of tryptophan are extremely potent mutagens. These compounds may also occur in cooked beef;[114] it has been stated, however, that the amounts are very small.[28] It is questionable whether this also holds for meat that has been in direct contact with fire, a situation occurring from time to time when barbecueing meat. It is still difficult to establish the health hazard caused by ingestion of PAH-containing foods, but it is a matter of fact that the contribution of industrially smoked meat and meat products to this problem is negligibly small.

4.2. Nitrosamines

Traces of nitrosamines may be present in meat products preserved with nitrite. These include mainly nitrosodimethylamine (NDMA), which may be present in amounts up to 1–2 μg/kg of meat product,[98] and nitrosopyrrolidine (NOPyr), which is developed during the frying of products like nitrite-cured bacon, and which may reach levels in the order of 10–20 μg/kg.[48] The nitrosamine problem with respect to meat products was reviewed a few years ago by Cassens *et al.*,[13] by Gray and Randall[48] and, in this series, by Walters.[158] Since then no essentially new knowledge concerning this matter has been gained; therefore, no further details are considered here.

4.3. Disinfectants

Disinfectants are widely used in the meat industry before handling and processing meat in order to improve hygienic conditions. Disinfectants that are not surface-active can be completely removed from tools and equipment by rinsing with water. Surface-active agents, however, adhere strongly to most surfaces and cannot be completely removed by simple rinsing. This holds especially for quaternary ammonium compounds. Amphotensids can be removed for the greater part, but not completely, in this way.[135]

When meat comes in contact with surfaces disinfected by these compounds, it will take up part of the adsorbed material, thus causing residue levels in the products varying from a few μg to nearly 1 mg/kg. The amounts ingested by consumption of meat and meat products are low in comparison to the daily amount taken up from drinking water, tableware and other products of civilisation, which ranges from 0·5 to 2·0 mg.[31]

The toxicity of these surface-active components is generally low,[31,135] but absorption of some nutrients by the intestinal tract may be partially inhibited. Their proper use for disinfection purposes in the meat industry does not introduce toxicological problems at the moment, but it should be taken into consideration that the ingestion of surface-active compounds by the consumer is tending to increase; therefore, more attention might be given in future to this problem. As for disinfectants that are not surface-active, e.g. chlorine and chlorine-releasing compounds, it should be stressed that these have to be carefully removed from disinfected surfaces, as free chlorine may react with several meat components, thus yielding substances some of which have toxic properties.[23]

Spices added during the manufacture of meat products are another possible source of contamination. These spices may have been disinfected by exposure to ethylene oxide or methyl bromide. Traces of these gases may adhere to the product; moreover, ethylene oxide may react with chlorine ions to form ethylene chlorohydrin (2-chloroethanol). By the addition of spices to certain types of sausages these components may contaminate the product. However, these could not be detected in sausages prepared with spices disinfected in this way.[71]

4.4. Compounds Migrating from Packaging Materials into Meat and Meat Products

Whilst glass, tinplate and paper have been widely used as food-packaging materials for many years, plastics are a relatively recent innovation in the packaging of food, including meat and meat products.

Although in themselves generally inert, plastics usually contain a variety of low-molecular weight compounds such as plasticisers, stabilisers, lubricants, unreacted or partially reacted monomer and impurities.[20] Most of these have lipophilic properties and tend to migrate from the packaging material into the fatty parts of the food packed. Not all these compounds are harmful to the consumer, but transfer of considerable amounts of these chemicals from the packaging material to the food should, of course, be avoided. Therefore, in several countries limits have been set for this migration. Some countries limit the amount of compounds subject to migration that are present in plastic films for food packaging, while other countries limit the amount of matter migrating into the product only. Modern legislation in the field of food packaging not only covers plastic packaging material but also the more 'classic' materials such as paper, rubber products, metals, glass and ceramics, textile products, etc.

Fresh meat is packed in order to protect it against external dirt and to allow hygienic handling, to retard moisture loss and to maintain a bright-red surface colour. For the last-mentioned purpose, ingress of oxygen is essential to convert the purplish muscle tissue to bright red oxyhaemoglobin. PVC and cellophane films are used for this purpose.[21] It should be noted that pure PVC is one of the more rigid plastics; therefore plasticisers, e.g. phthalate and adipate esters, are incorporated to obtain a flexible film. The content of these low-molecular plasticisers is 10–25% of the total weight. Because of the migration of these esters into the meat, this type of PVC film is unsuitable for packaging meat or meat products that have to be kept for longer than a few days. It should be noted in this respect that high-molecular plasticisers, e.g. polypropylene adipate, show only a slight tendency to migrate.

For longer-term storage, films impermeable to oxygen and water vapour are required. Copolymers of vinylidene chloride and vinyl chloride are used since these are suitable for vacuum packaging. Because of their toxicity some monomers require special attention. This holds particularly for the strongly carcinogenic vinyl chloride, which should not be present in the polymer in amounts exceeding 1 mg/kg, while the level in food products should be below 0·01 mg/kg.[30]

Finally, it has to be mentioned that contamination of meat with metals, as a result of packaging, should not be excluded. In an Italian study[102] an average value of lower than 0·05 mg Pb per kg was found for fresh beef, while in frozen beef samples the mean value was 0·16 mg/kg and for canned beef 0·79 mg/kg.

5. BIOCONTAMINANTS

Biocontaminants are substances produced by any living organism that are present in food and that are unwanted for some reason, mostly because of their toxic properties. This definition also includes compounds formed in food products as a result of microbial spoilage but, as stated earlier, these will not be considered here.

5.1. Mycotoxins

Mycotoxins are found on many plant products, particularly oilseeds and cereals. These substances are characterised by pronounced acute and chronic toxic properties, while some of them are potent carcinogens. Their production is governed by the temperature and water activity (a_w) of the substrate. Mycotoxin production generally does not occur below an a_w value of 0·83–0·85, while the temperature conditions vary with the toxin under consideration.[86] Although it must be noted that the mature animal is an effective toxin modifier and eliminator,[140] residues of mycotoxins are still present in animal tissues. Some metabolites should also be regarded as suspicious compounds, whose levels require careful monitoring. Among the many mycotoxins that may be present in minute amounts in meat and meat products, only the aflatoxins, ochratoxin A, T-2 toxin, zearalenone and penicillic acid will be discussed briefly in this section.

The group of *aflatoxins* consists of 17 closely related compounds that can be chemically classified as coumarin derivatives and that are produced mainly by *Aspergillus flavus* and *Asp. parasiticus*.

In 1961 it was found that the so-called 'turkey X disease', which caused a loss of 100 000 turkey poults in England, was due to the presence of a highly toxic substance in the groundnut meal with which the animals were fed. The toxin was obtained in crystalline form from the material and was found to be strongly hepatotoxic and carcinogenic. Rats fed contaminated groundnut meal or crystalline toxin developed hepatocarcinomas within a short time.[57] As the toxin present in the groundnut meal was shown to be produced by *Asp. flavus*, it received the name aflatoxin. Production is optimal at 27 °C and minimal at 12 °C.[112] The structures of some of the aflatoxins are as follows:

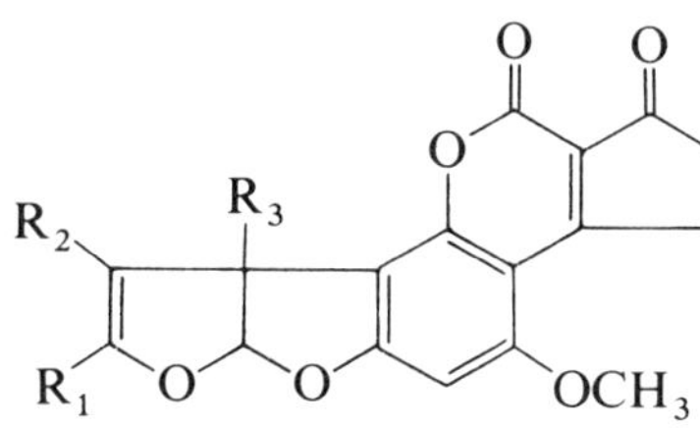

Aflatoxins of the B series

	R_1	R_2	R_3
B_1	H	H	H
B_2	H_2	H_2	H
M_1	H	H	OH
M_2	H_2	H_2	OH
B_{2a}	OH	H_2	H

The carcinogenic potential of B_1 is the highest and, as far as is known, unsurpassed in this respect by any other substance, either natural or synthetic.

Aflatoxin B_1 is readily metabolised in the animal body, the major metabolites being M_1 and B_{2a}. Aflatoxin B_{2a} is reported to be relatively non-toxic.[29] The carcinogenic potential of M_1 is only small as compared to that of B_1; however, aflatoxin M_1 should still be considered as strongly carcinogenic. Both B_1 and M_1 may exist in edible tissues of animals.[43,131,141] The feed to tissue ratio of B_1 for muscle tissue is in the order of 10^{-3}.[44,92] The highest concentrations are found in the liver. It has been stated[91] that aflatoxin B_1 can be detected in the livers of swine, broilers and cattle when levels in feed exceed 80, 100 and 1400 μg/kg, respectively. Pigs administered aflatoxin B_1 orally for 3 weeks in concentrations up to 1250 μg/kg of diet did not show residues of the unchanged compound at one day after withdrawal, while tissues were completely free of both B_1 and M_1 after 4 days.[111] The limit of detection for these components was 0·1 μg/kg. Similar results were obtained in another experiment.[45]

Tolerances have been set up in several countries for both food and feeds, mostly as the total concentrations of aflatoxins B_1, B_2, G_1 and G_2. Most limits are set up for all types of food and are in the range of 1–10 μg/kg. No special attention has been paid anywhere to meat and meat products.[137]

Like many other mycotoxins, aflatoxins are stable and heat-resistant compounds. Processing pork did not result in appreciable losses of aflatoxin B_1 and B_2.[44] It is possible that aflatoxin B_1 in animals is covalently linked to other compounds, thereby losing practically all of its carcinogenic properties.[66]

In contrast to the aflatoxins, which are formed primarily in tropical areas, the *ochratoxins* are generated in temperate climates as well, in particular on cereals. Production is possible at a temperature as low as 4 °C.[86] Ochratoxins are produced by several *Aspergillus* and *Penicillium* species and occur frequently in nature. At present, seven ochratoxins are known; however, only ochratoxin A is important from a public health standpoint. The molecule consists of a coumarin derivative linked to phenylalanine by a peptide bond.

Ochratoxin A (OA) has been thoroughly studied by Krogh.[83–85] It is strongly nephrotoxic; a high incidence of nephropathy in pigs is correlated with appreciable amounts of OA in barley, oats, maize and wheat. Pigs are particularly susceptible to this toxin. In the kidneys of pigs suffering from OA nephropathy 2–100 μg of OA per kg of kidney tissue was found and it was also present in liver, adipose and muscular tissue.

Ochratoxin A is a heat-stable substance, as demonstrated by its presence in roasted coffee and in beer.[103] Frying of products from swine reduced the amount of OA by 0–35% (mean 20%), while no losses were found in adipose tissue.[70] Ochratoxicosis in slaughter animals seems to be found mainly in Scandinavia, Yugoslavia and the USA. This is partly due to the use of home-grown cereals in these countries. In Denmark, limits have now been set up for OA.[137]

Human health hazards with respect to OA cannot be excluded, as human nephropathy is endemic in the Yugoslavian area where ochratoxicosis in slaughter pigs frequently occurs.[140] In ruminants, OA is easily hydrolysed by the action of proteolytic enzymes.[140] The possibility of OA residues in ruminant tissues can therefore be discounted.

T-2 toxin is a mycotoxin belonging to the group of the trichothecenes, a family of sesquiterpenoids. These compounds are cytotoxic to several kinds of animal and fungal cells. Anti-tumour activities have also been demonstrated.[152] T-2 toxin is produced by *Fusarium* species and by some other mould species as well. It is highly dermatotoxic and is able to cause necrosis of the mucous layers of the gastrointestinal tract. Severe oral lesions were observed when young chicks consumed feed with more than 1 mg/kg.[167]

Symptoms of T-2 intoxication have been observed in cattle, swine, poultry and possibly also in man.[92] Carry-over has been demonstrated in bovine and porcine milk[129] and in tissues of swine;[130] however, in the latter case residue concentrations are reported to be low. Further research is needed to establish whether T-2 toxin is a real health hazard to man.

Zearalenone, a mycotoxin frequently found in mould-infected maize, possesses oestrogenic properties and can be the cause of 'false heat' and infertility in swine herds. Teratogenic properties have also been reported. Like T-2 toxin, zearalenone is produced by some *Fusarium* species. *F. roseum* (*Gibberella zeae*) produces zearalenone in great abundance (3–15 g/kg[105]). Some carry-over to the milk has been observed in lactating cows[106] and to the liver in broilers,[107] but the amounts found are quite small. Zearalenone is partly metabolised to α-zearalenol, which is three times as oestrogenic as the parent component.[106]

Penicillic acid is a mycotoxin with reported cytotoxic, cardiotoxic and carcinogenic activity and which may occur in marked concentrations in maize. In comparison with aflatoxin B_1, penicillic acid has minimal toxicity to the chicken; its potential importance as a public health hazard, however, should not be excluded.[50] On the other hand, penicillic acid is considerably reactive towards SH groups of peptides and is therefore no longer detected shortly after ingestion.[92]

Proper control of mycotoxins in animal feedstuff is the basis for preventing contamination in food of animal origin.[47] Measures to achieve this include good farm management practices, good drying and storage conditions, shipping and handling procedures and good processing. In some cases the use of antifungal agents is helpful. It is sometimes not possible to prevent contamination completely and removal of mycotoxins may be taken into consideration. Treatments have been tested for the removal of aflatoxins, in particular. Extraction with appropriate solvents is effective in some cases. Treatment with ammonia and formaldehyde appears to be a promising chemical method, but effective processing differs markedly with the type of feedstuff. Another possibility which has recently been described is boiling in a salt solution.[35]

5.2. Phytotoxins

In several plants eaten by cattle, toxic compounds are present which may exert various effects on the animals. Isoflavones present in some locally abundant clover species are a typical example. These compounds are unwanted because of their oestrogenic properties.[19,138] Carry-over to the tissues is likely to be small, although traces have been observed in the livers of cattle.[49]

6. FINAL REMARKS

In general, it cannot be said that meat and meat products can be listed among the heavily contaminated foodstuffs. Environmental contamination is noticeable but not problematic, but contamination in game deserves some attention. It is obviously necessary to assess regularly the levels of polychlorinated biphenyls, heavy metals such as lead and especially cadmium, and some organochlorine pesticides as well. For that reason, monitoring programmes should be carried out.

The term 'monitoring' deserves some explanation. It was defined in 1973 by the United Nations Interagency Working Group on Monitoring as 'the process of repetitive observing, for defined purposes of one or more elements or indicators of the environment according to prearranged schedules in space and time and using comparable methodologies for environmental sensing and data collection. Monitoring provides factual information concerning the present state and the past trends in environmental behaviour' (cited from ref. 17). It is obvious that monitoring is extended to components present in the environment as a whole and not solely to foodstuffs. Nevertheless, the term can also be applied to the continuous checking for environmental contaminants in meat.

The processing of meat does not give rise to great concern at present. In industrially produced smoked meat products, carcinogenic polycyclic hydrocarbons do not play any role of significance. The level of harmful nitrosamines in nitrite-cured meat products, although already very low, will be further decreased in the future. It should be recognised, however, that other risks may be discovered as a result of our increasing knowledge.

As for residues of drugs in meat products, it is clear that some of them are unwanted at any concentration while others are acceptable at a certain level. In the first case the only good solution is a total prohibition of the use of such substances. In the second case, observance of withdrawal times may ensure that residues are indeed below the tolerance levels. It cannot be denied that unwanted residues are sometimes present or that the amount of other residues may exceed the level of tolerance. Obviously, these problems should not be attacked by monitoring but by effective supervision, which may also include screening programmes while the animals are still alive and not solely after slaughtering.

For this supervision, it is important that rapid methods that allow the economic handling of large numbers of samples become available. The development of immunoassays that can be performed in large series may be useful in this respect; because of the great selectivity of these methods, however, only one compound can usually be detected.

Another way of tackling the problem is the development of 'multi' methods in which more compounds will respond. It is true that methods in which inhibition of microbial growth is measured are responsive to a number of compounds used in veterinary practice, but these methods are slow and, for estimating residues of these compounds, the sensitivity is insufficient. Moreover, most metabolites cannot be detected in this way. Therefore, chemical and physical methods have to be applied for this purpose. A quite important step forward will be the development of techniques that allow a rapid clean-up of extracts from samples under investigation, to which the compounds to be analysed are transferred completely or almost completely and are concentrated to some extent. Such techniques are now under development and may contribute to effective methods for detection, estimation or determination of residues in meat and other products of animal origin in future.

Finally, the presence of traces of mycotoxins in meat is also not a reason for great concern. However, both monitoring and surveillance programmes remain necessary for an overview of the situation. As stated earlier, it is important that improvement of feed production and storage methods is encouraged, thus ensuring that the growth of toxin-producing moulds on feed or feed components is prevented.

7. REFERENCES

1. Adir, J. (1974). *Fed. Proc.*, **33**, 500.
2. Adir, J. (1975). *J. Pharm. Sci.*, **64**, 1847.
3. Aghte, O. and Dickel, H. (1980). *Arch. Lebensm. Hyg.*, **31**, 169.
4. Andelmann, J. B. and Suess, M. J. (1970). *Bull. WHO*, **43**, 479; ref. Suess, M. J., *Sci. Total Environ.*, **6** (1976) 239.
5. Anon. (1983). *Tijdschr. Diergeneesk.*, **108**, 584.
6. Barke, E., Hapke, H.-J., Heeschen, W., Kreuzer, W., Kübler, W. and Schreiber, W. (1983). *Rückstände in Lebensmitteln tierischer Herkunft: Situation und Beurteilung*, Verlag Chemie, Weinheim.
7. Benraad, Th. J., Stephany, R. W., Rosmalen, F. M. A., Hofman, J. A., Loeber, J. G. and Elvers, L. H. (1981). *Vet. Quart.*, **3**, 153.
8. Boiselle, C. and Langner, H. J. (1981). *Die Fleischerei*, **32**(4), I–IV; ref. Brunn, H., *Fleischwirtsch.*, **62** (1982) 1178.
9. Brabander, H. F. De and Verbeke, R. (1975). *J. Chromatog.*, **108**, 141.
10. Brooks, G. T. (1975). *Chlorinated insecticides*, Vol. II: *Biological and environmental aspects*, CRC Press, Cleveland, Ohio, p. 79.
11. Bruin, A. De and Zielhuis, R. L. (1971). *Tijdschr. Soc. Geneesk.*, **49**, 855.
12. Brunn, H. (1982). *Fleischwirtsch.*, **62**, 1178.
13. Cassens, R. G., Greaser, M. L., Ito, T. and Lee, M. (1979). *Fd Technol.*, **33**(7), 46.
14. Chambers, J. (1980). *Residue Rev.*, **73**, 101.
15. Cherain, M. G. (1979). *Environ. Health Perspect.*, **28**, 127.
16. Coordinating Committee for the Monitoring of Radioactive and Xenobiotic Substances (CCRX) (1982). *General radioactive contamination of the biosphere: Measurements in the Netherlands 1981*, Ministry of Public Health and Environmental Hygiene, No. 46.
17. Coordinating Committee for the Monitoring of Radioactive and Xenobiotic Substances (CCRX) (1982). *Measurements of xenobiotic substances in the biological environment in the Netherlands 1980–1981*, Ministry of Public Health and Environmental Hygiene, No. 69.
18. Corstiaensen, G. P., Bruin, J. J. M. De, Lendfers, L. H. H. M., Logtestijn, J. G. van and Verdijk, A. T. (1977). *Tijdschr. Diergeneesk.*, **102**, 811.
19. Cox, R. I. and Braden, A. W. (1974). *Proc. Aust. Soc. Animal Prod.*, **10**, 122.
20. Crompton, T. R. (1979). *Additive migration from plastics into food*, Pergamon Press, Oxford.

21. CROSBY, N. T. (1981). *Food packaging materials: Aspects of analysis and migration of contaminants*, Applied Science Publishers, London.
22. CRÖSSMANN, G. (1981). *Arch. Lebensmittelhyg.*, **32**, 87.
23. CUNNINGHAM, H. M. and LAWRENCE, G. A. (1977). *J. Fd Sci.*, **42**, 1504.
24. DERKS, H. J. G. M., FREUDENTHAL, J., LITJENS, J. L. M., GRAMBERG, L. G. and BORRIAS-VAN TONGEREN, V. (1983). *Biomed. Mass Spectrom.*, **10**, 209.
25. DEVLOO, S. (1972). *Tierärztl. Umschau*, **27**, 302.
26. DOREMIRE, M. E., HARMON, G. E. and PRATT, D. E. (1979). *J. Fd Sci.*, **44**, 622.
27. DOROUGH, H. W. (1970). *J. Agr. Fd Chem.*, **18**, 1015.
28. DUNN, B. P. (1982). Polycyclic aromatic hydrocarbons. In: *Carcinogens and mutagens in the environment*, Vol. I: *Food products*, CRC Press, Boca Raton, Florida, pp. 175–83.
29. DUTTON, M. F. and HEATHCOTE, J. G. (1968). *Chem. & Ind.*, 418.
30. EC Directive 78/142, 30 January 1978.
31. EDELMEYER, H. (1982). *Fleischwirtsch.*, **62**, 880.
32. EIKELENBOOM, G. (1981). Research and application of the halothane test in Dutch landrace pigs. *Proc. Symp. on Porcine Stress and Meat Quality* (Frøystein, T., Slinde, E. and Standal, N. (Eds)), Agricultural Food Research Society, Ås, Norway, pp. 260–7.
33. EISNER, H. J. and WULF, R. J. (1963). *J. Pharmacol. Exp. Ther.*, **142**, 122.
34. ENGST, R., MACHOLZ, R. M. and KUJAWA, M. (1974). *Nahrung*, **18**, 737.
35. FARAH, Z., MARTINS, M. J. R. and BACHMANN, M. R. (1983). *Lebensm. Wiss. Technol.*, **16**, 122.
36. FASSBENDER, C. P., RAMIREZ ALVAREZ, A. and WENZEL, S. (1977). *Arch. Lebensmittelhyg.*, **28**, 201.
37. FASSETT, D. W. (1975). *Ann. Rev. Pharmacol.*, **15**, 425.
38. FEINMAN, S. E. and MATHESON, J. C. (1978). *Subtherapeutic antibacterial agents in animal feeds*, Draft Environmental Impact Statement, Bureau of Veterinary Medicine, US Food and Drug Administration.
39. FLANJAK, J. and LEE, H. Y. (1979). *J. Sci. Fd Agric.*, **30**, 503.
40. FORSCHNER, F. and WOLF, H. O. (1981). *Arch. Lebensmittelhyg.*, **32**, 97.
41. FOX, M. R. S. (1983). *Fed. Proc.*, **42**, 1726.
42. FRIBERG, L., PISCATOR, M., NORDBERG, G. and KJELLSTRÖM, T. (Eds) (1974). *Cadmium in the environment*, 2nd edn, CRC Press, Cleveland, Ohio.
43. FURTADO, R. M., PEARSON, A. M., HOGBERG, M. G. and MILLER, E. R. (1979). *J. Agric. Fd Chem.*, **27**, 1351.
44. FURTADO, R. M., PEARSON, A. M., GRAY, J. I., HOGBERG, M. G. and MILLER, E. R. (1981). *J. Fd Sci.*, **46**, 1306.
45. FURTADO, R. M., PEARSON, A. M., HOGBERG, M. G., MILLER, E. R., GRAY, J. I. and AUST, S. D. (1982). *J. Agric. Fd Chem.*, **30**, 101.
46. GLAZKO, A. J. (1966). *Antimicrob. Ag. Chemother.*, 655; ref. *Chem. Abstr.*, **68** (1968) 28161 f.
47. GOLDBLATT, L. A. and DOLLEAR, F. G. (1979). Modifying mycotoxin contamination in feeds: use of mold inhibitors, ammoniating, roasting. *Proc. Symp. Interactions of Mycotoxins in Animal Production*, National Academy of Sciences, Washington, DC, p. 167.
48. GRAY, J. I. and RANDALL, C. J. (1979). *J. Fd Protect.*, **42**, 168.
49. GÜNTHER, H. O. (1978). *Fresenius' Z. Anal. Chem.*, **290**, 389.

50. Hanna, G. D., Phillips, T. D., Kubena, L. F., Cysewski, S. J., Ivie, G. W., Heidelbaugh, N. D., Witzel, D. A. and Hayes, A. W. (1981). *Poultry Sci.*, **60**, 2246.
51. Hansen, L. G. (1979). *Ann. NY Acad. Sci.*, **320**, 238.
52. Hansen, L. G., Tuinstra, L. G. M. Th., Kan, C. A., Strik, J. J. T. W. A. and Koeman, J. H. (1983). *J. Agric. Fd Chem.*, **31**, 254.
53. Hecht, H. (1980). *Fleischwirtsch.*, **60**, 1806.
54. Hecht, H. (1983). *Fleischwirtsch.*, **63**, 544.
55. Hecht, H., Schramel, P., Moreth, F. and Schinner, W. (1981). *Fleischwirtsch.*, **61**, 1316.
56. Henricks, D. M., Gray, S. L. and Hoover, J. L. B. (1983). Residue levels of endogenous estrogens in beef tissues. In: *Anabolics in animal production*, Proc. OIE Symposium, Paris (Meissonnier, E. (Ed.)), Office International des Epizooties, Paris, p. 233.
57. Hesseltine, C. W. (1979). Introduction, definition, and history of mycotoxins of importance to animal production. *Proc. Symp. Interactions of Mycotoxins in Animal Production*, National Academy of Sciences, Washington, DC, p. 3.
58. Heykants, J., Pardoel, L. and Janssen, A. A. J. (1971). *Arzneimittel-Forsch.*, **91**, 982.
59. Heykants, J., Symoens, J. and Marsboom, R. (1971). *Arzneimittel-Forsch.*, **21**, 1357.
60. Holm, J. (1978). *Fleischwirtsch.*, **58**, 1545.
61. Holm, J. (1979). *Fleischwirtsch.*, **59**, 737.
62. Holm, J. (1979). *Fleischwirtsch.*, **59**, 1345.
63. Holm, J. (1980). *Fleischwirtsch.*, **60**, 2227.
64. Holm, J. and Elsarnagawy, D. (1980). *Fleischwirtsch.*, **60**, 1387.
65. International Agency for Research on Cancer (1983). *IARC Monographs*, **31**, 13, 39, 147.
66. Jaggi, W., Lutz, W. K., Lüthy, W., Zweifel, U. and Schlatter, C. (1981). *Fd Cosmet. Toxicol.*, **18**, 257.
67. Jansen, E. H. J. M., Stephany, E. W., Vos, J. G. and Ruitenberg, E. J. (1983). *Ber. RIV*, **20**, 160.
68. Janski, A. (1983). Development of a sensitive method for extraction and assay of zeranol residues in animal tissues and the use of the method in a multiple implant study in cattle. In: *Anabolics in animal production*, Proc. OIE Symposium, Paris (Meissonnier, E. (Ed.)), Office International des Epizooties, Paris, p. 443.
69. Joint Expert Committee on Food Additives (JECFA) (1981). *Evaluation of certain food additives*, 25th Report, Technical Rep. Ser. 669. World Health Organisation, Geneva, 48 pp.
70. Josefsson, B. G. E. and Möller, T. E. (1980). *J. Sci. Fd Agric.*, **31**, 1313.
71. Jordy, A. (1981). *Fleischwirtsch.*, **61**, 1667.
72. Käferstein, F. K., Altmann, H.-J., Kallischnigg, G., Klein, H., Kossen, M.-T., Lorenz, H., Müller, J., Schmidt, E. and Zufelde, K. P. (1979). *Blei, Cadmium und Quecksilber in und auf Lebensmitteln: Modellstudie*, ZEBS-Berichte 1/1979, Bundesgesundheitsamt, Berlin.
73. Kan, C. A. (1978). *J. Agric. Fd Chem.*, **26**, 1051.

74. KAN, C. A. and JONKER-DEN ROOYEN, J. C. (1978). *J. Agric. Fd Chem.*, **26**, 935.
75. KELLY, R. G. and BUYSKE, D. A. (1980). *J. Pharmacol. Exp. Ther.*, **130**, 144.
76. KLINE, R. D., HAYS, V. W. and CROMWELL, G. L. (1972). *J. Animal Sci.*, **34**, 393.
77. KNÖPPLER, H.-O., GRAUNKE, W., MÜCKE, W., SCHULZE, H. and GEDEK, W. (1979). *Fleischwirtsch.*, **59**, 241.
78. KOH, T. S., BENSON, T. H. and JUDSON, G. J. (1980). *J. Assoc. Off. Anal. Chem.*, **63**, 809.
79. KONERMANN, H., CRÖSSMANN, G. and HOPPENBROCK, K. H. (1982). *Tierärztl. Umschau*, **37**(8), 11.
80. KREUZER, W., BUNZL, K. and KRACKE, W. (1979). *Fleischwirtsch.*, **59**, 1529.
81. KREUZER, W., BUNZL, K. and KRACKE, W. (1981). *Fleischwirtsch.*, **61**, 1886.
82. KREUZER, W. and ROSOPULO, A. (1981). *Arch. Lebensmittelhyg.*, **32**, 181.
83. KROGH, P., HALD, B., HASSELAGER, F., MADSEN, A., MORTENSEN, H. P., LARSEN, A. E. and CAMPBELL, A. D. (1973). *Pure and Appl. Chem.*, **35**, 275.
84. KROGH, P., ELLING, F., HALD, B., LARSEN, A. E., LILLEHØJ, E. B., MADSEN, A. and MORTENSEN, H. P. (1976). *Toxicology*, **6**, 235.
85. KROGH, P. (1977). Ochratoxins. In: *Mycotoxins in human and animal health* (Rodricks, J. V., Hesseltine, C. W. and Mehlman, M. A. (Eds)), Pathotox Publishers, Park Forest South, Illinois, p. 189.
86. KROGH, P. (1980). Carry-over of mycotoxins into poultry products. *Proc. 2nd European Symposium on Poultry Nutrition*, Beekbergen, The Netherlands, 1979 (Kan, C. A. and Simons, P. C. M. (Eds)), PUDOC, Wageningen, p. 88.
87. KRZEMINSKI, L. F., COX, B. L. and GOSLINE, R. E. (1981). *J. Agric. Fd Chem.*, **29**, 387.
88. LAITEM, L. and GASPAR, P. (1977). *J. Chromatog.*, **140**, 266.
89. LARSSON, B. K., SAHLBERG, G. P., ERIKSSON, A. Y. and BUSK, L. A. (1981). *J. Agric. Fd Chem.*, **31**, 867.
90. LEE, J. J. VAN DER (1981). Determination of chloramphenicol and its application to residues in milk and in dairy cows. Thesis, Leiden, p. 188.
91. LEISTNER, L. and HOFMANN, G. (1979). Carry-over von Mycotoxinen in Fleisch. *Symp. Carry-over von Schadstoffen in der landwirtschaftlichen und tierischen Produktion.* Ref. *Fleischwirtsch.*, **60** (1980) 1904.
92. LEISTNER, L. and ECKARDT, C. (1981). Schimmelpilze und Mycotoxine in Fleisch und Fleischerzeugnissen. In: *Mycotoxine in Lebensmitteln* (Reiss., J. (Ed.)), Gustav Fischer Verlag, Stuttgart/New York, p. 297.
93. LINDEMAYR, H., KNÓBLER, R., KRAFT, D. and BAUMGARTNER, W. (1981). *Allergy*, **36**, 471.
94. LISKA, B. J., STEMP, A. R. and STADELMAN, W. J. (1967). *Fd Technol.*, **21**, 435.
95. LISTER, D., GREGORY, N. G. and WARRIS, P. D. (1981). Stress in meat animals. In: *Developments in meat science—2* (Lawrie, R. (Ed.)), Applied Science Publishers, London, p. 61.
96. LO, M.-T. and SANDI, E. (1978). *Residue Rev.*, **69**, 35.
97. LOGTEN, M. J. VAN, LEEUWEN, F. X. R. VAN and STEPHANY, R. W. (1981). *Tijdschr. Diergeneesk.*, **106**, 353.
98. LÖTZSCH, R. (1979). *Fleischwirtsch.*, **59**, 683.
99. LU, C., MCCALLA, A. R. and BRYANT, D. W. (1979). *Mutation Res.*, **67**, 133.

100. LYINSKY, W. and ROSS, A. E. (1967). *Fd Cosmet. Toxicol.*, **5**, 343.
101. MADARENA, G., DAZZI, G., CAMPANINI, G. and MAGGI, E. (1980). *Meat Sci.*, **4**, 157.
102. MAGGI, E., BRACCHI, P. C., CAMPANINI, G., DAZZI, G. and MADARENA, G. (1979). *Meat Sci.*, **3**, 309.
103. MASSART, D. L., DEELSTRA, H. and HOOGEWIJS, G. (1980). *Vreemde stoffen in onze voeding*, De Nederlandsche Boekhandel, Antwerp/Amsterdam, p. 155.
104. MIRNA, A. and CORETTI, K. (1979). *Meat Sci.*, **3**, 97.
105. MIROCHA, C. J., PATHRE, S. V. and CHRISTENSEN, C. M. (1977). Zearalenone. In: *Mycotoxins in human and animal health* (Rodricks, J. V., Hesseltine, C. W. and Mehlman, M. A. (Eds)), Pathotox Publishers, Park Forest South, Illinois, p. 345.
106. MIROCHA, C. J., PATHRE, S. V. and ROBISON, T. S. (1981). *Fd Cosmet. Toxicol.*, **19**, 25.
107. MIROCHA, C. J., ROBISON, T. S., PAWLOSKY, R. J. and ALLEN, N. K. (1982). *Toxicol. & Appl. Pharmacol.*, **66**, 77.
108. MORETH, F. and HECHT, H. (1981). *Fleischwirtsch.*, **61**, 1326.
109. MORGAN, K. J., ZABIK, M. E. and FUNK, K. (1972). *Poultry Sci.*, **51**, 470.
110. NEFF, A. W. (1983). Analytical methods for melengesterol acetate®. In: *Anabolics in animal production*, Proc. OIE Symposium, Paris (Meissonnier, E. (Ed.)), Office International des Epizooties, Paris, p. 457.
111. NEFF, G. L. and EDDS, G. T. (1981). *Fd Cosmet. Toxicol.*, **19**, 739.
112. NORTHOLT, M. D., EGMOND, H. P. VAN and PAULSCH, W. E. (1977). *J. Fd Protect.*, **40**, 778.
113. OLLING, M., STEPHANY, R. W. and RAUWS, A. G. (1981). *J. Vet. Pharmacol. Therap.*, **4**, 291.
114. PARIZA, M. W. (1982). *Fd Technol.*, **36**(3), 53.
115. PERIĆ, M., RAŠETA, J., VIŠACKI, V. and SPIRIĆ, A. (1980). *Technologija Mesa*, **21**, 132.
116. POHLSCHMIDT, J. (1979). *Fleischwirtsch.*, **59**, 253.
117. POTTHAST, K. (1978). *Fleischwirtsch.*, **58**, 69.
118. POTTHAST, K. (1978). *Fleischwirtsch.*, **58**, 340.
119. POTTHAST, K. (1979). *Fleischwirtsch.*, **59**, 1515.
120. POTTHAST, K. (1980). *Fleischwirtsch.*, **60**, 1941.
121. PREISS, A. (1983). *Fleischwirtsch.*, **63**, 243.
122. RAUWS, A. G., OLLING, M., FREUDENTHAL, J. and HAM, M. TEN (1976). *Toxicol. & Appl. Pharmacol.*, **35**, 333.
123. RAUWS, A. G. and OLLING, M. (1978). *J. Vet. Pharmacol. Therap.*, **1**, 57.
124. RAUWS, A. G. (1983). *Tijdschr. Diergeneesk.*, **108**, 659.
125. REED, D. L., BOOTH, N. H., RUSH, P. B., GOETSCH, D. D. and KIKER, J. (1977). *Poultry Sci.*, **56**, 908.
126. RHEE, K. S. and BRATZLER, L. J. (1970). *J. Fd Sci.*, **35**, 146.
127. RITCHEY, S. J., YOUNG, R. W. and ESSARY, E. O. (1969). *J. Fd Sci.*, **34**, 169.
128. RITCHEY, S. J., YOUNG, R. W. and ESSARY, E. O. (1972). *J. Agric. Fd Chem.*, **20**, 291.
129. ROBISON, T. S., MIROCHA, C. J., KURTZ, H. J., BEHRENS, J. C., CHI, M. S., WEAVER, G. A. and NYSTROM, S. D. (1979). *J. Dairy Sci.*, **62**, 637.

130. Robison, T. S., Mirocha, C. J., Kurtz, H. J., Behrens, J. C., Weaver, G. A. and Chi, M. S. (1979). *J. Agric. Fd Chem.*, **27**, 1411.
131. Rodricks, J. V. and Stoloff, L. (1977). Aflatoxin residues from contaminated feed in edible tissues of food-producing animals. In: *Mycotoxins in human and animal health* (Rodricks, J. V., Hesseltine, C. W. and Mehlman, M. A. (Eds)), Pathotox Publishers, Park Forest South, Illinois, p. 67.
132. Ryan, J. J. and Hoffmann, B. (1978). *J. Assoc. Off. Anal. Chem.*, **61**, 1274.
133. Salmi, A. and Hirn, J. (1981). *Fleischwirtsch.*, **61**, 1175.
134. Sanders, W. E. and Sanders, C. C. (1979). *Ann. Rev. Pharmacol. Toxicol.*, **19**, 53.
135. Schmidt, U. (1983). *Fleischwirtsch.*, **63**, 227.
136. Schoffer, D. and Hoffmann, B. (1981). *Arch. Lebensm. Hyg.*, **32**, 141.
137. Schuller, P. L., Stoloff, L. and Egmond, H. P. van (1983). Limits and regulations on mycotoxins. *Proc. Int. Symp. on Mycotoxins*, Cairo, 1981 (Naguib, K. *et al.* (Eds)), National Information and Documentation Centre, Cairo, p. 111.
138. Shutt, D. A. (1976). *Endeavour*, **35**, 110.
139. Staarink, T. and Hakkenbrak, P. (1982). *Het Contaminantenboekje* (survey on levels of contaminants in food products from the Dutch market), Staatsuitgeverij, The Hague.
140. Stoloff, L. (1979). Mycotoxin residues in edible animal tissues. *Proc. Symp. Interactions of Mycotoxins in Animal Production*, National Academy of Sciences, Washington, DC, p. 157.
141. Stoloff, L. (1983). *J. Assoc. Off. Anal. Chem.*, **66**, 355.
142. Sümmermann, W., Rohleder, H. and Korte, F. (1978). *Z. Lebensm.-Unters. Forsch.*, **166**, 137.
143. Tazima, Y., Kada, T. and Murakami, A. (1975). *Mutation Res.*, **32**, 55.
144. Thorsteinsson, Th. (1969). *Cancer*, **23**, 455.
145. Tilgner, D. J. (1968). *Fd Manuf.*, **43**(6), 37.
146. Tóth, L. (1971). *Fleischwirtsch.*, **51**, 1069.
147. Tóth, L. and Blaas, W. (1972). *Fleischwirtsch.*, **52**, 1121.
148. Tóth, L. and Blaas, W. (1972). *Fleischwirtsch.*, **52**, 1419.
149. Trottmann, K. and Blumenthal, A. (1982). *Mitt. Geb. Lebensm. Hyg.*, **73**, 208.
150. Tschudi, P. (1973). *Zbl. Veterinärmed. A*, **20**, 145, 155.
151. Tuinstra, L. G. M. Th., Vreman, K., Roos, A. H. and Keukens, H. J. (1981). *Neth. Milk & Dairy J.*, **35**, 147.
152. Ueno, Y. (1977). Trichothecenes: overview address. In: *Mycotoxins in human and animal health* (Rodricks, J. V., Hesseltine, C. W. and Mehlman, M. A. (Eds)), Pathotox Publishers, Park Forest South, Illinois, p. 189.
153. United Nations Scientific Committee on the Effects of Atomic Radiation (UNSCEAR) (1982). *Ionizing radiation: sources and biological effects*, Annex B, par. 85–88, Table 17.
154. Verschuuren, H. G., Wit, S. L., Kamp, C. G. van de, Vries, Th. de and Logten, M. J. van (1973). Report No. 1/73 Tox, National Institute for Public Health, Bilthoven, The Netherlands.

155. VOS, J. G., STEPHANY, R. W., CASPERS, J. W., LOON, J. TH. G. VAN, METZLAR, J. W. H. and OVERHAUS, H. B. M. (1982). *Vet. Quart.*, **4**, 1.
156. VREMAN, K., TUINSTRA, L. G. M. TH. and POORTVLIET, L. J. (1978). *Proc. 24th Conf. of European Meat Research Workers*, Kulmbach, L 7:1.
157. WAGNER, J. F. (1983). Estradiol controlled release implants: efficacy and drug delivery. In: *Anabolics in animal production*, Proc. OIE Symposium, Paris (Meissonnier, E. (Ed.)), Office International des Epizooties, Paris, p. 129.
158. WALTERS, C. L. (1980). Nitrosamines in meat products. In: *Developments in meat science—1* (Lawrie, R. (Ed.)), Applied Science Publishers, London, p. 195.
159. WEBB, M. and ETIENNE, A. T. (1977). *Biochem. Pharmacol.*, **26**, 25.
160. WEINSTEIN, L. (1975). Chemotherapy of microbial diseases. In *The pharmacological basis of therapeutics* (Goodman, L. S. and Gilman, A. (Eds)), Macmillan, New York, p. 1090.
161. WIT, S. L. and KAMP, C. G. VAN DE (1973). Report No. 47/73 Tox-RoB, National Institute for Public Health, Bilthoven, The Netherlands.
162. WOHLBIER, W. and SCHNEIDER, W. (1966). *Z. Tierphysiol., Tierernähr. u. Futterm.*, **21**, 34.
163. WOLFORD, S. T. and ARGOUDELIS, C. J. (1979). *J. Dairy Sci.*, **62**, 1458.
164. WORLD HEALTH ORGANIZATION (WHO) (1972). *Evaluation of mercury, lead, cadmium and the food additives amaranth, diethyl pyrocarbonate and octyl gallate*, WHO Food Additive Ser. No. 4.
165. WORLD HEALTH ORGANIZATION (WHO) (1972). *Evaluation of certain food additives and the contaminants mercury, lead and cadmium.* WHO Technical Report Ser. No. 505.
166. WRIGHT, F. C. and RINER, J. C. (1979). *J. Agric. Fd Chem.*, **27**, 576.
167. WYATT, R. D., WEEKS, B. A., HAMILTON, P. B. and BURMEISTER, H. R. (1972). *Appl. Microbiol.*, **24**, 251.
168. ZMUDZKI, J. (1977). *Medycyna Weterinaryjna*, **33**, 41; ref. *Fd Sci. & Technol. Abstr.*, **9** (1977) S1778.

INDEX